US Defense Strategy from to Operation Iraqi Freedom

The book examines the thirty-year transformation in American military thought and practice that spanned from American military withdrawal from Vietnam through the 2003 invasion of Iraq. During these three decades, new technology and operational practices helped form what observers dubbed an American "Revolution in Military Affairs" in the 1990s and a "new American way of war" in the 2000s. Drawing on a diverse range of recently declassified documents, interviews, innovation research, and personal experience, Tomes tells for the first time the story of how innovative approaches to solving battlefield challenges gave rise to nonnuclear strategic strike, the quest to apply information technology to offset Soviet military advantages, and the rise of "decisive operations" in American strategy. The historical chapters provide the first serious consideration of the evolution of military capabilities and doctrine that underwrote a rapid dominance approach to military operations and recent preemption language in US national security strategy.

Tomes provides historical context for understanding the post-Vietnam renewal in American military affairs and the key military innovations of the 1970s and 1980s by reviewing the evolution of Cold War national military strategy. He documents how capabilities designed to defend NATO evolved into a training revolution, precision strike, stealth aircraft, joint doctrine, and integrated intelligence capabilities; these are the core elements of current US military dominance. Among the contributions to strategic studies is the book's exploration of how research and development strategies conceived in the late 1970s influenced later research and development activities. An important contribution to military innovation studies, the book suggests an innovation framework applicable to the study of both past and current defense transformations.

This book will be of great interest to all students of US military thought and defense planning, to students of strategic studies, and to those interested in the general history of military affairs since the Second World War.

Robert R. Tomes is Chief of the Strategic Initiatives Group in the Analysis and Production Directorate of the National Geospatial-Intelligence Agency, a member of the Council for Emerging National Security Affairs, and a Director of the Anna Sobol Levy Foundation.

Strategy and History
Series Editors: Colin Gray and Williamson Murray
ISSN: 1473–6403

This new series will focus on the theory and practice of strategy. Following Clausewitz, strategy has been understood to mean the use made of force, and the threat of the use of force, for the ends of policy. This series is as interested in ideas as in historical cases of grand strategy and military strategy in action. All historical periods, near and past, and even future, are of interest. In addition to original monographs, the series will from time to time publish edited reprints of neglected classics as well as collections of essays.

This book offers a very thoughtful and sophisticated analysis of the processes and prospects of defense transformation. Drawing on the experience of the Cold War and the aftermath of the Vietnam war, it brings the reader right up to the current debates about "revolutions in military affairs" and the Bush Doctrine on preemption. Extremely well-written and carefully researched, it will be valuable reading for anyone interested in defense policy.

(Professor George Quester, University of Maryland)

Essential reading in order to understand, from a military, technological, and geopolitical perspective, the dramatic transformation on US defense posture from the end of World War II to the present "war on terrorism."

(Jacques S. Gansler, Vice President Research, University of Maryland, former Under Secretary of Defense (1997–2001))

If you really want to understand what has happened to the US armed forces since Vietnam, read this book.

(Professor Martin Van Creveld)

This conceptually grounded book provides an historical account of the thirty-year journey toward the modern US military that is fighting in Iraq, Afghanistan and hotspots around the globe.

(Dr Peter Dombrowski, Chair, Strategic Research Department, US Naval War College)

US Defense Strategy from Vietnam to Operation Iraqi Freedom

Military innovation and the new
American way of war, 1973–2003

Robert R. Tomes

Routledge
Taylor & Francis Group

LONDON AND NEW YORK

First published 2007
by Routledge
2 Park Square, Milton Park, Abingdon, Oxon OX14 4RN

Simultaneously published in the USA and Canada
by Routledge
270 Madison Ave, New York, NY 10016

*Routledge is an imprint of the Taylor & Francis Group,
an informa business*

© 2007 Robert R. Tomes

Typeset in Times New Roman by
Newgen Imaging Systems (P) Ltd, Chennai, India
Printed and bound in Great Britain by
Biddles Digital, King's Lynn

British Library Cataloguing in Publication Data
A catalogue record for this book is available from the British Library

Library of Congress Cataloging in Publication Data
A catalog record for this book has been requested

ISBN10: 0–415–77074–2 (hbk)
ISBN10: 0–415–77252–4 (pbk)
ISBN10: 0–203–96841–7 (ebk)

ISBN13: 978–0–415–77074–3 (hbk)
ISBN13: 978–0–415–77252–5 (pbk)
ISBN13: 978–0–203–96841–3 (ebk)

Contents

Illustrations

Figures

Table

Acknowledgments

This is not the only book that will be written about the origins of the new American way of war. But it is among the first works to call for a more deliberate understanding of the origins of current military forces and the approach to military operations that is now driving military history. My intent was to provide a useful departure point for understanding military innovation and the origins of the new American way of war.

No book is written in a vacuum. I was lucky to have help, encouragement, and most of all constructive criticism. Anything readers find useful or informative in the following pages I credit to others, reserving for myself sole credit for mistakes, omissions, and any failure to get the story right. Some of those that offered assistance or advice along the way will not agree with the final draft; others will lament that I've left out their favorite parts of the story. To this I can only offer the consolation that others with better story-telling skills will surely write on the same period of American military innovation.

Several people played key roles in the evolution of this study. George Quester at the University of Maryland, College Park, shaped my initial thinking about military innovation and provided sage advice through years of sometimes frustrating research. He helped me find my voice. Peter Dombrowski, now at the Naval War College, provided support and friendship that I probably do not deserve. For over a decade I've relied on his advice and candor. He read and commented on successive drafts. Richard Van Atta of the Institute for Defense Analyses commented on key chapters, providing critical material on DARPA and the offset strategy. Without his help, key parts of the story would be missing. Readers seeking additional information on the offset strategy or DARPA's many Cold War successes should read his excellent case studies.

Warren Phillips and Dave Lalman also provided important feedback. Warren opened his rolodex to help me find additional interviewees. Dave asked insightful questions about military innovation processes. Invaluable comments on earlier drafts of the manuscript or much earlier drafts outlining my arguments about military change were provided by Steve Canby, ADM William Crowe, and James Wirtz. Kevin Cunningham, former Dean of the Army War College, graciously and patiently suffered through many phone calls to discuss American military strategy during the Cold War. His invitations to discuss innovation with his seminar students helped me refine my argument. Had he survived pancreatic cancer, I would have been honored for him to provide a forward to this book.

A number of people granted interviews. The Honorable Jacques Gansler provided important insights into the world of defense transformation, including personal accounts of his role championing GPS in the early days of this remarkable program. Loren Larson shared crucial insights and some key documents from his time in the Defense Department's Conventional Initiatives Office. He shared his knowledge of the origins of precision strike and the Assault Breaker program. LTG James King shared personal thoughts on post-Vietnam military innovation. Lt. Gen. James Clapper's candid observations about intelligence reform were invaluable. LTG Pat Hughes, affectionately known as "Yoda" to a generation of Army intelligence officers, is indeed a master of his profession. His personal accounts of what the Vietnam generation experienced were sobering. King, Clapper, and Hughes provided important insights into the evolution of military intelligence and the travails of leading reform. They personalized parts of the story that later generations will always struggle to understand.

LTG Stan McChrystal offered a unique perspective on the current generation of Army leaders' ability to innovate on the battlefield. Major General Mike Pfister and Peter Oleson clarified aspects of the historical period discussed in chapters four and five. Andrew Krepinevich focused my thinking on military revolutions. John Young shared his experiences coordinating national intelligence estimates on the Soviet Union during the latter stages of the Cold War. Chris Haakon and Rich Johnson provided insights into the evolution of geodesy and mapping. Pete Rustan helped me understand the obstacles to innovation that exist within most government agencies and how passionate people find ways to make a difference. Other interviewees asked to remain anonymous; to all I'm indebted.

My research and writing has been supported, informed, and improved by association with defense and intelligence professionals that graciously discussed and debated key aspects of the story or provided much-needed encouragement. Rob Zitz, Robert Cardillo, Keith Masback, Ed Mornston, Jeff Mayo, and Bill Wansley all patiently suffered through discussions of military innovation theory and then pointed out how the real world works. Zitz's masterful ability to tell the story of precision strike inspired me. Kevin O'Connell helped with funding to finish research while I was at RAND. Lisa Witzig and Bryan Maizlish challenged my thoughts on innovation. Winston Beauchamp, Michele Weslander, Jim Seybold, and Dennis Miller supported and encouraged my research while I was at the National Geospatial-Intelligence Agency.

Deborah Barger kept me grounded and challenged my thinking during her time as an Intelligence Fellow at RAND. Her own work on innovation and transformation deserves a wider audience. VADM John Morgan did not read the manuscript, but his willingness to debate the principles of war from the perspective of a battle group commander sharpened my views on warfare in the information age. Bruce Berkowitz, my neighbor at RAND, provided encouragement and his own unique views on the changing nature of warfare. Martin Van Creveld did not read the manuscript but unknowingly provided much needed advice and encouragement at critical times over the last decade; to him I owe my understanding of what Nietzsche meant by setbacks making us stronger.

Finally, I'm indebted to Susan Holley, who patiently endured months of research and writing. Without her love, devotion, and compassion, none of this would have been possible. She's the one I want in my foxhole.

This book is dedicated to my parents, who always knew.

1 Military innovation and defense transformation

American military strategy is in flux. It is too early to tell how the experience of Operation Iraqi Freedom will influence military doctrine, training, operational approaches to future small wars, or even force structure. It is also too early to tell how protean changes in post-9/11 American national security, including shifts in defense transformation, will influence the further evolution of American military strategy. What seems clear is that the period of military transformation that has been underway since Vietnam has come to an end.

The 2003 invasion of Iraq is likely to be the last conflict to showcase what has been called a new American way of war, what in the 1990s was dubbed an American revolution in military affairs. Reversing late Cold War trends, military strategists, defense planners, and doctrine writers are focusing on small wars, counterinsurgencies, and nation-building operations—missions that the US defense planning community largely ignored during the past three decades. This book focuses on the evolution of defense planning and military strategy from the end of US military involvement in Vietnam to the invasion of Iraq, documents how military innovation processes led to American military dominance over other conventional militaries, and suggests how the contextual and organizational aspects of defense planning during this period focused efforts away from counterinsurgency warfare. In some respects it is a book on why hard lessons from counterinsurgency warfare in the 1960s and 1970s were not learned in the 1980s and 1990s.

Military strategy is the art and process of conceptualizing the forms of military power required of national security or defense strategy, managing resources to provide military power, and applying military power effectively to fulfill ends outlined in national strategy. At a more basic level, strategy involves the instrumental relationship between purpose and power, a relationship defined in part by decisions to allocate resources. Such decisions are influenced by attitudes, biases, expectations, and a host of contextual factors that influence what resources are available when, where, and for how long. Attitudes and resources, furthermore, are always reciprocally related, with the former shaping the realm of the possible and latter the realm of the probable. Military strategy writ large is often caught in the tension between these two realms, largely because attitudinal or conceptual factors like doctrine affect the actual military effectiveness of an army as much as more tangible resource measures such as manpower or numbers of tanks.

And sometimes doctrinal innovations lead to increased military effectiveness without any changes in forces or manpower.

An unexplored aspect of military effectiveness, one that is often at the roots of successful military strategies, is the role military innovation plays in resolving the tension between attitudes and resources, between the expectations for military power across all mission areas and the capabilities available to meet them.

Recognizing that much uncertainty exists in defining future threats, and lacking a crystal ball through which to discern the types of battles American forces must prepare to fight decades from now, this study falls back on an old adage: we live life forward but understand it backwards, through history. The affinity for smaller, lighter, more lethal, networked forces that was first chartered in the 1980s is being reinforced; this trend has not been fully evaluated against the capabilities needed for the war on terrorism.

Questions about the ongoing transformation in American military capabilities remain. How and why did rapid deployment forces and rapid dominance concepts ascend in American national military strategy? What about widespread arguments for increasing the size of US special operations forces and developing urban warfare capabilities? Few works assess changes in Special Forces and urban warfare in light of other developments, including changing beliefs about the role of military force in international relations after the Cold War. Too few studies, moreover, place the origins of emphasis on information and decision superiority in the context of the transition from a Cold War defense of Europe to regional conflicts against far weaker opponents.

In the midst of often free-wheeling discussions about American defense transformation, one is reminded of the downside of the American cultural antipathy toward knowing our own history. Defense analysts, lamentably, have not yet placed the lessons of recent conflicts into a larger historical context to surface historic themes and patterns likely to dominate warfare in the coming decades.

This study reviews historical, conceptual, and doctrinal factors central to the evolution of US defense policy and military thought over the course of three decades, with emphasis on the maturation of what was dubbed a revolution in military affairs (RMA) in the 1990s and a new American way of war in the 2000s.

What constitutes an RMA? RMAs usually involve a major shift in the nature of warfare brought about by innovative applications of new technologies in combination with fundamental changes in doctrine, operational practices, and organizations. Usually identified in hindsight, after a stunning military success, RMAs involve radical changes in the conduct of military operations and sometimes even the characterization of warfighting. RMA theory and definitions are addressed in a later chapter. It suffices here to mention that the 1990s witnessed a shift in American military thought and defense discourse as new terms and concepts were widely used to describe US military forces, doctrine, and capabilities.

Soviet military theorists were the first to identify new American military capabilities as exhibiting RMA-like changes. Soviet writers actually coined the term RMA in the 1950s to describe changes in warfare wrought by nuclear weapons and ballistic missiles. One of the misconceptions this book addresses is the tendency for some analysts to date RMA discussions much later–the 1970s or early 1980s, overlooking the origins and longevity of the term and associated concepts about how warfare was changing. More importantly, many have overlooked how some of the same changes in warfare attributed to nuclear weapons were replicated by advanced conventional forces.

This is an important part of Cold War military history that is too easily overlooked. In identifying American military capabilities with an RMA, Soviet and then US military analysts were communicating something profound about the historical importance of US long-range precision strike capabilities, which were replicating the battlefield effects small nuclear weapons had on armored forces. In the current era of American preponderance, other states are seeking to offset US military advantages by designing around the capabilities the United States developed in the last decades of the Cold War.

American military innovation during the Cold War largely involved responses to three related sources of strategic and operational necessity: attempts to correct or stabilize imbalances in the nuclear deterrence equation; challenges in peripheral regions that had the potential to escalate into a crisis; and specific operational threats to US or NATO forces that had strategic implications for East–West stability. By the 1980s, security challenges in each area called for advanced conventional warfighting forces.

Conventional warfighting innovations were pursued to restore deterrence credibility in Europe. Even as US defense planners sought to address concerns about the military balance in Europe, Soviet expansionism into the Third World raised additional concerns about the ability to contain a regional conflict and prevent escalation to a wider global one. Anxiety about Soviet penetration into strategically important regions like the Persian Gulf was based in part on the emergence of new strategic and operational challenges to American military forces. Responses to these challenges evolved throughout the late 1970s and early 1980s and led to a new arsenal of nonnuclear strategic strike capabilities. These were the capabilities Soviet observers labeled an RMA.

A wellspring of studies and prolific media references to 'revolutionary' warfighting capabilities permeated defense planning discussions following the American military victory over Iraq in the 1991 Gulf War. There was often more rhetoric than reality in many of the pro-RMA arguments of the time. Even among those more conservative analysts choosing not to directly evoke RMA terminology, defense planning discussions and military thought were dominated by what this study terms an American *RMA thesis*. This thesis faded in the late 1990s but remains central to more recent defense transformation discussions. The American RMA thesis holds that a historically significant shift in US military power was underway by the end of the Cold War based on the synergy of advanced intelligence, surveillance, and reconnaissance (ISR) capabilities, automated target identification systems, information-enabled weapons, superior education and training, and joint warfighting capabilities.

In the 2000s, many of the concepts and ideas associated with the RMA are being folded into defense transformation discussions. Among the RMA terminology retained in US defense discourse are terms like information superiority, rapid dominance, dominant battlespace knowledge, common operating pictures, decision superiority, persistent surveillance, and full spectrum dominance. The same changes in military effectiveness that analysts dubbed revolutionary in the 1990s are underwriting shifts in national security strategy, including the mention of preemptive military strikes within national security planning documents.

Increased references to preemption in US national strategy in the early 2000s, arguably, reflected both unprecedented confidence in American military power and a realization that the previous mainstays of doctrine—deterrence, containment, and engagement—were not adequate for a global war on terrorism. The references to preemptive strikes in American national security strategy reflect, in part, the effect of decades of arguments among strategists about nonnuclear precision strike capabilities. Since the 1970s and 1980s, strategists have argued that advanced conventional forces would lower the threshold for using military force as a foreign policy tool.

Some argued positively, suggesting too many limits were placed on the use of military force as a tool to coerce rogue states. Defense luminaries like Albert Wohlstetter enjoined in debate throughout the 1970s and 1980s, leading to the 1987 report of the Commission on Integrated Long-Term Strategy. The commission, co-chaired by Wohlstetter, remains an important milestone in the evolution of military strategy and provides insights into the intellectual roots of the 2002 *National Security Strategy* and its 2006 revision.

This study explores the relationship between American military dominance, or at least its origins in Cold War times, and more recent changes in national security strategy that, some argue, reflect increased willingness to use military force to achieve national policy objectives (an argument that Wohlstetter and others made throughout the 1980s). Later chapters consider how successive, successful military operations amplified an affinity for a rapid dominance approach to warfighting and galvanized optimistic expectations—and perhaps overconfidence—about the efficacy of using military force to achieve foreign policy objectives.

Examining the antecedents to the new American way of war provides perspective on how the evolution of military capabilities and doctrine during the last decades of the Cold War continue to influence American strategic culture, defense transformation, and approaches to military operations. The story of how military forces developed in the 1970s and 1980s to restore deterrence stability is also the story of how Cold War military innovations came to underwrite a preemption strategy in the early 2000s; this is a fascinating, largely unknown story of technological, doctrinal, and operational innovation.

Study overview

Following the 1945 London Conference of Foreign Ministers, with the Grand Alliance disintegrating, as diplomats grappled to prevent an escalation of the emerging Cold War, and while statesmen struggled to understand the implications of the dawning nuclear era, American and Soviet defense strategy proceeded along different evolutionary paths. Whereas American defense strategy evolved around the US nuclear monopoly and soon focused on nuclear weapons and their delivery means, Soviet defense planners had no choice other than strengthening conventional warfighting capabilities.

In time, nuclear arsenals became the linchpin of Cold War deterrence stability. Nuclear strategy would be the cornerstone of both US and Soviet grand strategy by the end of the 1950s. American defense strategy, in turn, was by then dominated by nuclear weapons innovations, nuclear targeting strategy, and nuclear deterrence theory. Sustained quests to achieve, preserve, or reclaim deterrence stability in the face of asymmetric weapons developments drove American foreign policy and shaped the military doctrine proscribing how forces would fight.

In emphasizing nuclear strategy and nuclear warfighting capabilities, the United States all but ignored conventional strategy and innovations to modernize general purpose or conventional warfighting forces. Similarly, military thought relating to nonnuclear combat stagnated. American ground force modernization, to restate the point, was neglected or assigned low priority until relatively late in the Cold War era. The 1960s and 1970s did bring efforts to develop nonnuclear capabilities for Third World and peripheral contingencies, but on a very limited basis. Meaningful conventional force modernization did not transpire until the late 1970s and early 1980s.

Why at that point, after decades of relative neglect? Nonnuclear capabilities, quite simply, were suddenly deemed a strategically viable solution to credibility problems with the West's nuclear deterrent. It was this push toward conventional modernization, an attempt to bolster nuclear deterrence, that led to the capabilities observes dubbed an American RMA in the aftermath of the 1991 Gulf War.

By the early 2000s, nuclear strategy slipped to the margins of national security studies and all but disappeared from core public debates about defense planning. Indeed, the only noticeable public discussion of nuclear weapons concerned the proposed development of new, low-yield, tactical nuclear weapons to attack underground facilities and bunkers that conventional weapons could not destroy. This is the essence of the thirty-year transformation involving the post-Vietnam resurgence in American conventional military power.

The idea of a thirty-year transformation in American defense strategy has not been studied sufficiently. Indeed, many analysts have all but ignored the most important reversal at the heart of this decades-long transformation process. In the 1970s, at the beginning of the thirty-year transformation, conventional force modernization was pursued as an adjunct to nuclear deterrence and to restore perceived losses in deterrence stability; in the 2000s, partly in response to the challenges posed by underground facilities, a new class of "bunker-busting" nuclear weapons were proposed as an adjunct to global, nonnuclear precision strike capabilities. Later chapters explore additional changes associated with the shift from a defense strategy dominated by nulear strategy and doctrine to one dominated by advanced conventional forces, a rapid dominance approach to regional wars, and global information dominance.

Underscoring this study is the view that, after a decade of arguments and rhetoric about an ongoing RMA, too little analysis exists on the innovations *anteceding* the fielding of so-called "revolutionary" American military capabilities. In other words, there are few scholarly works placing the thirty-year transformation in context, a problem highlighted by growing references to a new American way of war.

But there is a larger problem with the paucity of works on the evolution of American military thought and defense strategy. It seems disingenuous and epistemologically unsound for students of US defense policy and military thought to proceed headlong into assessments of the current defense reform period without some degree of perspective—historically and intellectually—on the origins and evolution of current capabilities. Moreover, much of the RMA thesis that dominated 1990s defense planning discourse seems inappropriate as a guide to preparing the military for expanding the Global War on Terrorism or even the "stability operations" (e.g. peace keeping, counterinsurgency) defense planners now associate with defense transformation.

For sure, this is only one of many books required to tell the story of American military innovation, the story of military, technological, social, political, and other factors associated with the ascent of American conventional military power in the late twentieth century. Among the capabilities discussed in later chapters are long-range precision strike, stealth technology, air-ground operations, Global Positioning System applications, night vision capabilities to "own the night," realistic training, an all volunteer force, joint doctrine, the ascent of a distinctly American science of systems engineering, and the advent of a knowledge-based approach to operations.

Two objectives are pursued in this study. My first objective is to document key aspects of the thirty-year transformation in American military effectiveness that occurred from 1973 through 2003, with an emphasis on the 1970s and 1980s. A synthetic approach summarizes key themes and events, providing context and perspective to understand the thirty-year innovation process that began in the shadow of the US evacuation of Saigon, matured in the 1980s as Pentagon planners sought an integrated nuclear-conventional deterrent, was refined during the 1990s as American military forces were deployed to address regional crises, and culminated in 2003 with successful attacks through blinding sandstorms on the road to Baghdad.

My second objective is to use the story of the thirty-year transformation to derive a military innovation framework that others will hopefully find useful as a guide to explore other military innovation processes and to think through defense transformation processes.

Along the way I contend that a military innovation framework is the most appropriate conceptual approach for 1) understanding the origins of so-called "revolutionary" American military capabilities and 2) using this understanding to inform current or future defense transformation decisions. Innovation studies are well positioned to draw on the theoretical, historical, and policy dimensions of previous works on discontinuous changes in military effectiveness; they also provide ample room for focusing on continuities across periods of change. The wide and varied field of defense policy studies will benefit from additional frameworks that refract lessons of previous innovation cases through lenses attuned to today's strategic and operational challenges.

Chapter 2 surveys military innovation theory and proposes a framework for studying innovation processes. This is not an attempt to develop and prove a new theory explaining all cases of military innovation across all periods of military affairs. The objective is expanding military innovation literature for students of the political and military sciences. Chapter 2 argues that military innovation studies offer additional insights into ways to think about major military changes involving new, novel, and breakthrough changes by unpacking complex innovation processes into discrete and analyzable historical narratives. In doing so, they locate decision makers in past innovation milieus that may differ widely in scope and scale—but not necessarily in kind—from contemporary ones.

Chapter 3 reviews early Cold War American military thought and defense planning. The principle narrative of the thirty-year transformation is a relatively straightforward story that is best understood by first understanding the evolution of American military strategy from the Second World War through Vietnam. Reviewing Cold War nuclear strategy, particularly deterrence strategy and nuclear targeting developments, instills appreciation for later changes in American military doctrine and defense planning. These changes, traceable to developments in the 1970s and early 1980s, are best understood from the perspective of what was *overturned* in military thought and doctrine by the centering of the RMA thesis. Chapter 3 concludes that a dominant narrative of nuclear strategy in defense policy discourse, military thought, and doctrine evolved within American national security policy, one that constrained thinking about conventional warfighting capabilities (termed "general purpose" forces in contemporary writings).

Chapter 4 focuses on the origins of important innovations dating from the 1970s as the US military began its recovery from Vietnam. This was a transitional period in American national security. During the 1970s, inflation reached 14 percent. Gas rationing was imposed as long lines appeared at filling stations. Millions cancelled their vacations. A number of important events occurred in 1977, which marked something of a turning point in the history of conventional warfare. That year, then Secretary of Defense Harold Brown and William Perry, the Director of Defense Research and Engineering, launched a new research and development strategy known as the "offset strategy."[1]

Perry later argued that post–Cold War advances in US military effectiveness descended from this strategy, named for technologies (e.g. sensors, precision-guided weapons, and stealth technologies) that "would give qualitative advantages to American forces to offset the quantitative advantage the Soviet forces enjoyed."[2] It was these capabilities (including technologies) associated with the offset strategy that later "achieved the status of a 'revolution in military affairs.' "[3]

Also in 1977, the Defense Advanced Research Project Agency aligned its budget to address conventional theater challenges, including information systems applied to battlefield decision making, command and control, and surveillance systems. Lockheed flew a technology demonstration airplane leading to the F-117 stealth bomber, a plane that would never had been designed, built, or flown without computers. Signals for what became a space-based Global Positioning System (GPS) were proven adequate for ground navigation and maneuver. The Air Force's Airborne Warning and Control System (AWACS) and other battlefield remote sensing systems entered operational service and redefined the notion of tactical reconnaissance by enabling distributed theater awareness. A new space-borne reconnaissance capability for remote sensing entered operational use, eventually linking national (strategic) remote sensing and warning tools directly to military operations.

Shifting strategic realities and pressing operational challenges created a milieu ripe for multidimensional innovations. In 1979, with American hostages still in Iran, President Jimmy Carter kept the national Christmas tree dark–a symbolic move that defined the national mood. The Carter Doctrine extended US military power to defend the Persian Gulf from Soviet expansion; the failed April 1980 Iran hostage rescue mission spurred additional military reforms; and a presidential directive brought the largest US arms build-up in three decades.

During this time, the then classified Assault Breaker concept demonstration was created to develop weapons systems able to "rip the heart out of" the Soviet Army—to destroy its armored forces.[4] Assault Breaker was, arguably, part of a larger vision for using information technology in what today's defense analysts call a "system-of-systems"; it was the original plan for linking systems with other systems using information technology.

Chapter 5 covers the 1980s, focusing on the period leading up to the 1986 signing of the Goldwater-Nichols Act, the Army's publication of a revised AirLand Battle doctrine, and Soviet Premier Mikhail Gorbachev's consolidation of power in Moscow. During the 1980s, American military strategy diversified, losing its nuclear-centric focus. Conventional deterrence theory ascended in national security discussions. Joint warfighting was emphasized; training and doctrine were revamped. Information technology began to be viewed as the foundation of battlefield weapons systems and warfighting capabilities. Specific technological innovations included long-range precision strike capabilities drawing on GPS, theater reconnaissance assets, and information-enabled, integrated weapons platforms. US defense spending nearly tripled during this period.

In the early 1980s, defense analysts argued that Soviet conventional forces were becoming much more capable and therefore more of a threat to NATO forces. NATO's conventional defenses seemed less viable, less able to defend against an attack. Lacking a conventional defense and retaliation alternative, awakening to the post-Vietnam imperative for radical reform, and succumbing to political pressure at home and abroad to reduce reliance on nuclear deterrence, US military planners undertook a series of initiatives that coalesced into a significant increase in US military effectiveness.

A conceptual revolution occurred in part because concept demonstrations and experiments gave defense planners and military leaders a clear vision of how new weapons and battlefield sensing systems might resolve pressing operational challenges. A training revolution both reflected and reinforced professionalism as expectations about the efficacy of new technology evolved. Expectations for technology also affected doctrinal changes as technology and doctrine became more intertwined. A tripwire mentality relying on nuclear deterrence no longer assuaged Allies fearful that, after theater nuclear weapons were employed on Allied territory to blunt a Soviet advance, a cease-fire negotiation would trade Allied territory to prevent global nuclear war. For these and other reasons, conventional force proponents were ready to offer alternatives to nuclear weapons when the strategic and operational needs to do so became an imperative.

As the 1980s closed, the narrative of nuclear strategy no longer dominated defense discourse, key elements of the information-enabled precision reconnaissance-strike system were under development or already in service, and the post-Vietnam antipathy to engage militarity abroad seemed to be waning. A vision for the future of warfare, therefore, existed before the Cold War ended and the RMA thesis ascended in US defense planning.

Chapters 4 and 5 focus on Army and Air Force developments. For sure, the Marine Corps and Navy undertook innovations during this period as well, and the Navy was responsible for important developments in guidance systems and targeting. Navy work on network centric warfare was almost adopted wholesale into American military thought in the early 2000s. Still, Army and Air Force innovations figured more prominently in the origins and evolution of the American military capabilities underwriting the new American way of war. These capabilities were conceived to address strategic and operational challenges in the European theater. As Vice Admiral (retired) Bill Owens argues, the advent of "new technology and a shift toward different operational concepts" in the 1970s "was most prominent in the U.S. Army"; the Army "began to develop a much greater capacity to see and track events at greater distances and attack with longer-range, precision weapons."[5]

Chapter 6 reviews post–Cold War defense policy discourse, including a critical review of the so-called RMA debate. It examines the effects of the end of the Cold War on American military thought and reviews key aspects of post–Cold War defense strategy that relate to the thirty-year transformation in American defense strategy. Chapter 6 also sketches the role of the information revolution in the evolution of the American RMA and on current defense transformation visions. Chapter 7 concludes the study, revisiting the military innovation framework discussed in Chapter 2 from the perspective of historical information presented in Chapters 4, 5, and 6.

2 On military innovation

The decades-long evolution of American military capabilities, doctrine, and mindset involves an understudied and often overlooked story of military innovation.

Many scholars have explored why, how, and to what end nations make major innovations in the way they organize, equip, and employ military forces.[1] Historians and political scientists analyze military innovations to describe why and how they take root, to theorize about their manifestation in the form of operational capabilities or altered security relations, and to assess how civil, military, or other factors found within innovation processes affect outcomes. Some aim to develop or bolster new or existing theoretical frameworks to explain the conditions accounting for successful innovations; a few do so to specifically suggest how these conditions might be replicated.[2]

This chapter reviews concepts, theories, and literature related to military innovation studies. We explore several questions. What is innovation? What types of studies inform the military innovation subfield within political science? Are current theories and cases on US military innovation positioned to inform ongoing transformation discussions? If not, what revisions or innovation cases are required? The intent is sketching, with broad strokes, the outlines of innovation as a phenomenon distinct academically, epistemologically, and organizationally from other behavior.

Introducing military innovation studies

Military innovation studies are fundamentally and epistemologically about understanding and describing qualitative improvements in *military* effectiveness that yield a comparative advantage over other militaries, creating opportunities for increasing a nation's overall *strategic* effectiveness. For Colin S. Gray, strategic effectiveness involves the "the net (i.e. with the adversary dimension factored in) effectiveness of grand strategic performance, which is to say of behavior relevant to the threat or actual use of force."[3]

Although many elements contribute to military innovation, those associated with significant advances in grand strategic effectiveness often trace their origins to necessity wrought from some mix of strategic, operational, or tactical challenges that limit military organizations from mastering an existing core competency or a new mission area. These challenges can be immediate, in the form of an existing battlefield problem, or perceived to be a challenge because of a new problem.

Why should we care about the serious exploration of past cases of military innovation? Military innovations are of growing interest to scholars concerned with war studies, power transitions, and a myriad of other international security issues involving qualitative shifts in force correlations between or among competitors. Such shifts draw interdisciplinary interest because they relate to other aspects of global politics and security. They also illuminate how specific organizational and political processes affect later security arrangements and war outcomes. Likewise, because they provide fertile ground for theory building and hypothesis testing, innovation cases are employed to polish theoretical lenses attuned to issues as diverse as threat perception, offensive–defensive theory, deterrence theory, arms control, and technology diffusion.

Barry Watts and Williamson Murray view the "underlying purpose" for innovation studies as "helping decision makers to think creatively about changes in the nature of war that may occur in coming decades," not merely examining "historical episodes for their own sakes."[4] Understanding the ebbs and flows of previous innovations, Murray argues elsewhere, illuminates "how military institutions innovate" in generalized terms, which for contemporary policy makers suggests how innovation and transformation initiatives might alter "performance on the battlefields of the twenty-first century."[5] We are also concerned with the antecedent conditions and other factors that provide context within which military innovations succeed or fail. These include how understanding of current or future threats lead to proposals for change.

As military historian Allan Millet observes, the "essence of justifiable innovation" stems from "strategic calculation and the analysis of perceived threat."[6] Williamson Murray similarly concludes that, although not an absolute in every case, "one precondition for significant military innovation" seems to be "a concrete problem which the military institutions involved have vital interests in solving."[7] These views of innovation are borne out in Chapters 4 and 5, which discuss strategic and operational challenges facing US forces in Central Europe. This does not mean that scrutinizing innovation activity is the sole source of insight into significant advances in combat power available to analysts. To paraphrase Eugene Gholz, however, they are "a crucial independent variable in good theories of victory."[8]

Although different degrees of innovation exist, what frequently matters to those interested in engendering a discontinuous increase in strategic effectiveness are significant military innovations that diverge from standard practices or prevailing ways of warfare. They often involve some mix of untried, disruptive technological, operational, and organizational changes.

Major military innovations or major periods of innovation in military affairs are about large-scale, historically noteworthy change that over time shifts military effectiveness. Differentiating between lesser innovations and historically noteworthy ones is difficult given the widely disparate contextual factors subsuming innovation activities across time, cultures, and socio-technical domains. A single theoretical bent is unlikely to capture the richness of the underlying behavior.

Analytically, it remains useful to differentiate historically momentous examples of military innovation from the more routine march of military science. Even major military innovations are not necessarily harbingers of military revolution or revolutions in military affairs, although neither seems possible without one or more innovations. Perhaps it is more appropriate to argue that major innovations are necessary but not sufficient for the emergence of a revolution in military affairs (RMA).

Like warfare itself, military innovation is a social activity hinging, in this case, upon pockets of cooperative behavior (often in the face of stern opposition) aiming to alter organizational missions and activities in response to some strategic or operational need. This frequently occurs to correct a real or perceived, existential or anticipated, specific or general, performance gap that has strategic implications. During periods of "great historical challenges" or "at times of crisis," military historian and philosopher Azar Gat opines, new ideas emerge expressing "human effort to come to grips with new developments and integrate them within meaningful intellectual frameworks."[9]

Within an organization, innovators may perceive themselves as zealots on a mission, viewing their work as saving the organization by redefining identities and mapping an organization's core values to emerging or future operational realities. Success, from a leadership perspective, depends on some mix of disciples, champions, and organizational discipline. Innovations, then, usually require some type of external or high-level sponsorship to achieve successful implementation and diffusion. Innovations also tend to disrupt organizations by causing a change in business processes or a change in how the organization measures strategic effectiveness. Examples include German *Stormtrooper* tactics, the development of amphibious landing capabilities in the US Navy and Marine Corps, the Air Force's turn to long-range strategic bombing after the Munich Pact, and Israel's development of an offensive doctrine.

Coming to terms with military innovation

Innovation definitions vary. Organizational theorists consider innovation to be simply "the creation and implementation of a new idea" as "long as the idea is perceived as new and entails novel change for the actors involved."[10] Political scientist and student of military doctrine Barry Posen defines innovation as a "large change" originating from organizational failure, external pressures, or an organization's expansionist policy.[11] James Q. Wilson views innovations as new programs or technologies that "involve the performance of new tasks or a significant alteration in the way in which existing tasks are performed."[12] For Wilson, "[r]eal innovations are those that alter core tasks," a conclusion that resonates in the realm of military innovation where Service roles and missions are indeed organized around key military tasks.[13] At the basic level, these tasks are divided into aerospace, land, sea, and littoral warfare domains.

Stephen Peter Rosen defines a "major innovation" in similar terms. They involve "a change that forces one of the primary combat arms of a service to change its concepts of operation and its relation to other combat arms, and to abandon or downgrade traditional missions." Overall, he concludes, significant military innovations "involve changes in critical tasks, the tasks around which warplans revolve."[14] This is an apt description of what happened in the thirty-year transformation giving rise to a new American way of war.

In the burgeoning business management literature on innovation, strategic and operational necessity relates to organizational performance, with innovation often driven by current or emerging performance gaps, market changes, or shifts in customer expectations.[15] Legendary management theorist Peter Drucker simply defines innovation as "change that creates a new dimension of performance."[16] In his classic *The Comparative Advantage of Nations*, Michael E. Porter sees innovation as an outcome of "unusual effort" to embolden "new or improved ways of competing" designed to overcome "pressure, necessity, or even adversity." For Porter, furthermore, "fear of loss often proves more powerful than the hope of gain," an insight applicable to the innovation period studies in later chapters.[17]

Drawing on Wilson, Rosen, and others, defining aspects of military innovation for this study include an understanding of an organization's core tasks, the relationship between tasks and war plans, changes in the strategic (or operational) environment, war plan viability given such changes, and how efficiently organizations accommodate adjustments in missions or tasks required by the new capability.

Innovation is not invention, although the invention of a thing, idea, or concept often antecedes the articulation and diffusion of an innovative application of technology or new approach to warfare. Invention—of a thing or an idea—is antecedent to innovation, which is the outcome of applied invention or inventions mixed with opportunity and the will to attempt change.

The key to successful innovation lies in the health and welfare of the diffusion and adoption process. In this study, innovation subsumes diffusion and adoption: there can be no successful innovation if the advantage proffered by the proposed new capability never enters service. Indeed, military innovation studies are partly about the efficacy of processes stewarding new capabilities with the potential to increase effectiveness. The anteceding discovery–invention process is often a solitary one, focused on the development of something new. The innovation, diffusion, adoption process is focused on maximizing the outcome of ideas, technology, or processes in terms of performance, either in an organization, on behavior, or within a market.

Innovations are applied within and through organizations to achieve significantly better or qualitatively different *military* outcomes. They involve some mix of opportunity and necessity; interests and values; calculated determination and sheer luck. Ultimately a teleological enterprise, successfully diffusing a major innovation involves disruption, displacement, and divergent thinking; innovators, by definition, rub against the accepted order of things and are exemplars of entrepreneurial activism.

Above all, innovations change organizational outcomes in terms of how resources are mustered to accomplish objectives and missions. Moreover, they change intra-organizational dynamics, sometimes bringing conflict between organizations, an important element of innovation behavior returned to in later chapters.

Contextualizing innovation

Context, as well as interdependent aspects of both structure and agency (or the behavior of specific actors), is a reoccurring theme in military innovation studies that warrants additional attention. Williamson Murray and MacGregor Knox, for example, contend that "revolutions in military affairs always occur within the context of politics and strategy—and that context is everything."[18] For political scientists in general, the strategic *and* socio-political environments together define the strategic and operational context. In this larger context, organizations and leaders define and pursue objectives, missions, and tasks. But in many studies context is too narrowly defined, often because the author is constrained by a theory attuned to domestic or organizational politics or one that favors changes in the international balance of power as the predominate factor in military change.

"Strategic innovation," which is really the domain of military innovations pursuant to transformational changes, is, as Richard Betts concludes, dependant "on the social and political milieu."[19] Drawing on Betts, this study evokes the term *innovation milieu* to describe the nexus of challenges and opportunities within which military innovations occur. Here, the innovation milieu framework subsumes interaction effects of both structure and agency, the primary elements of innovation systems, processes, and actors that exist in specific moments within specific organizational settings. A French term, a milieu is a point or coordinate in space and time that includes both the middle and its surroundings. Conceptually it has no beginning or end, more a nexus of connections, relationships, and potential influence pathways. In the world of innovation, it is a useful image to capture how organizations and systems interact with their environments— in both contextual and ideational terms—such that over time certain relationships influence paths and evolve profound, sometimes revolutionary, change.

Retired Admiral William Owens posits that new technology and concepts originating in the 1970s, many of which became central to the American RMA, appeared after "new generations of nuclear and conventional weapons required novel approaches by the Army and Air Force to maintain the credibility of deterrence in Europe."[20] "It was in this milieu," he continues, "that technologies and operational concepts arose that would be central to" what observers later called an American RMA.[21] Such military innovations as radar, the Norden bomb sight, amphibious assault technology and doctrine, the German blitzkrieg, nuclear weapons, and stealth technology emerged from specific strategic and operational milieus characterized by necessity and focused ingenuity.

In these cases and others, the innovation milieu took root in a context of strategic and operational challenges that pushed military or defense organizations toward new ideas and ways of accomplishing military tasks. In this sense, the innovation milieu construct follows Secretary Emeritus of the Smithsonian Institution Robert McC. Adams' observation that "innovations are better understood less as independent events that unleashed new sequences of change in their own right than as periodically emergent outcomes of wider, interactive systems."[22] Such interactive systems are the domain of military innovation studies.[23]

Generally, the contextual elements of an innovation milieu can be identified as specific areas of analysis or investigation, as suggested in Figure 2.1. Chapter 7 revisits this framework, using it to unpack major aspects of the history of American Military innovation during the last decades of the Cold War and to suggest how students of defense transformation may benefit from a reinvigoration of military innovation studies.

Murray discusses another important aspect of context. "Military innovations that have the greatest influence are those that change the context within which wars take place."[24] This suggests two sides to the analytic problem of contextualizing military innovations: understanding the contextual antecedents and conditions that shape an innovation milieu ripe for significant changes in military effectiveness and then understanding how military innovations themselves alter or

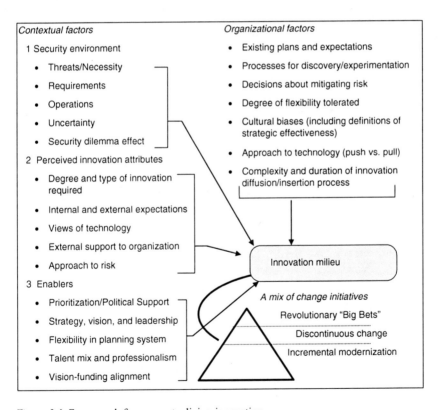

Figure 2.1 Framework for conceptualizing innovation.

otherwise influence military affairs. For policy makers pursuing transformative changes in their military services, it is often the details of the former that are of greatest import; for students of military history it is frequently the effect of military innovations that draws the greatest interest. Here, attention is given to both.

Contextual factors, to reinforce the point, are vital for the maturation and success of innovation, which "occur in organizational contexts that both enable and motivate innovation."[25] Conversely, of course, context can act as a barrier, preventing the coming together of an innovation milieu ripe for the type and degree of innovation behavior of interest to students of major military innovation.[26]

Although the theories and findings of business management scholars do not always lend themselves to military innovation studies, a point discussed later, their increasing focus on contextual factors within which significant change occurs, including case studies, illuminates the critical role of environmental factors in success or failure. Such factors range from perceptions, leadership support, the degree of urgency underlying pursuit of change, and qualities of the organization's culture.

Economist Nathan Rosenberg adds that innovation "involves extremely complex relations among sets of key variables," including "inventions, innovations, diffusion paths, and investment activity," and further concludes that "innovation and diffusion rates" are "powerfully shaped by expectation patterns."[27] Innovations, moreover, depend "upon an entire supporting infrastructure."[28] In other words, context counts, perception matters, and technology is not the final arbiter of technology innovation. His analysis also emphasizes the role of systems and social networks in innovation, both of which concern organizational expectations about missions and performance. Rosenberg differentiates between innovations and *major* innovations, the latter providing "a framework for a large number of subsequent innovations, each of which is dependent upon, or complementary to, the original one." They "constitute new building blocks which provide a basis for subsequent technologies."[29]

Students of military innovation explore four levels or areas of context. The first concerns the strategic environment in which an organization is nested, including its larger political setting, what this study refers to as strategic dynamics. The second are those factors intrinsic to how innovation is perceived in light of the security environment. This includes how the organization is socially and materially constructed as a functional unit serving some larger societal purpose (e.g. national defense). The third level of context involves what this study terms enablers, which includes resource alignment. Finally, innovation is contextualized within the organization itself as a discrete entity.

Although some argue for the primacy of variables common only to one or two of these levels, this study finds that all are important, as are interactions among levels—a point returned to in the conclusion. The process of translating strategic imperatives into required organizational capabilities (or outcomes) requires some understanding of the new measures of effectiveness required. Innovation scholars seeking to understand such processes cannot adequately assess innovation strategies, processes, or changes in effectiveness without first understanding the underlying rationale to pursue significant change.

Military innovation studies: a sketch of resources

What resources from traditional political and military science schools of thought inform military innovation studies? This section suggests four categories of works from history and political science that tend to be more accessible and familiar to students of defense analysis, military science, and international security. They inform this study of innovation. The first and second are primary and secondary works on innovation cases and periods, many from outside the formal discipline of political science; the preponderance of these are historical studies, essays, and memoirs. The third includes political science and international relations works addressing doctrinal innovations. In the fourth category are multifaceted approaches to military innovation and military change.

Collapsing a rich and diverse spectrum of works into categories is inherently problematic; some will surely object to the categories themselves. One benefit of providing some structure to the range of available sources is providing a more *inclusive* survey of works than that found in current studies, most of which attempt no review of the universe of existing resources. Indeed, they tend to review only those theories at the root of their specific take on innovation behavior. My intent is to point others to works that, regardless of their theoretical bent, expand awareness of the rich field. The categorization is not presented as *the* standard for others nor does it claim to be comprehensive.

This first category of studies helps sketch the topography and contextual nuances of military cultures, institutions, and other characteristics of military organizations through the prism of direct experience or association. As the broadest, it includes memoirs, autobiographies, and select analytical essays on the US defense establishment. In addition to conceptual and historical analysis, included here are essays and writings by military reform participants and policy makers that discuss, describe, or otherwise provide insight into decisions undertaken before and during periods of innovation or organizational change.

Carl Builder's *The Masks of War: American Military Styles in Strategy* remains a classic work on American military culture and Service identities. Builder informs military innovation studies by helping students of military organizations understand decisions concerning roles and missions, the procurement of new weapons systems and combat platforms, and the effect of organizational culture influences on innovation choices. Others include two biographies of John Boyd (one by Grant Hammond, the other by Robert Coram) that discuss the culture of defense reform in the 1970s, James Burton's memoir on defense reform, and Kenneth Adelman and Norman Augustine's analysis of technological and geopolitical influences on US defense policy.[30] These works provide insights into the culture of change inside defense organizations.

The second category explores military innovation and modernization processes, focusing on specific cases or technologies. These works tend to focus on simplifying the complexity and nuances of innovation cases to render innovation processes and outcomes more accessible to those seeking to understand the dynamics of military change. They do so, generally, as fairly straightforward narratives concerned with conveying insights along historical dimensions rather than through theories, frameworks, and policy-focused analysis.

Representative of these works are Robert Buderi's history of radar, aptly entitled *The Invention that Changed the World* and Harvey Sapolsky's *The Polaris System Development: Bureaucratic and Programmatic Success in Government.*[31] Other variants focus on historical processes that yield conclusions about the phenomena of military innovation for specific historical cases, often embedded within a particular context. Nicholas A. Lambert's *Sir John Fisher's Naval Revolution*, which documents innovations in British naval defense in the period before the First World War, carefully dissects the interaction among strategic and technological changes, organizational reforms, political pressures, and the role of leaders (e.g. Winston Churchill) in bringing doctrinal and other innovations to fruition.

Another multifaceted account is Frank Winter's history of rocket technology in the nineteenth century, *The First Golden Age of Rocketry*, which documents how innovations by William Congrave and William Hale affected military technology, whaling, torpedoes, and other areas. A classic study in this category is Elting E. Morrison's chapter on innovations in naval gunfire at sea in *Men, Machines and Modern Times*, which demonstrates that organizational identity and personality factors are sometimes more important to innovations than technological and doctrinal changes alone. Bruce Gudmundsson's *Stormtrooper Tactics: Innovation in the German Army, 1914–1918*, documents the German innovations in infantry tactics during the First World War that spurred the transformation of German military thought and doctrine. Gudmundsson attributes German innovation to the decentralized nature of German military organizations, a proclivity for self-education within German culture, an early start toward change compared to other nations, and innovations in operational art.

William Odom, in *After the Trenches: The Transformation of U.S. Army Doctrine, 1918–1939*, concludes that successful modernization requires "procurement of enough equipment for experimentation"; an adept foreign intelligence organization; and, "an organization dedicated to monitoring and accommodating change."[32] James Kitfield's *Prodigal Soldiers* reviews cultural, doctrinal, and technological factors central to the US Army's post-Vietnam renewal.[33]

Also included here are studies of specific systems not focused on innovation processes or innovative technologies per se. Richard Betts' edited volume *Cruise Missiles: Technologies, Strategy, Politics* is representative of similar works that touch on aspects of innovation in larger studies of weapons or technologies.[34] Allan R. Millet and Williamson Murray's three-volume series on military effectiveness are additional examples.[35] Among the noteworthy aspects of their edited volumes is a serious investigation of strategic measures of effectiveness. Although significant military innovation is not necessary to achieve superior military effectiveness, and although military effectiveness often increases when organizations perfect established procedures or technologies (not innovative ones), effectiveness studies remain important sources of insight into military change.

MacGregor Knox and Williamson Murray's edited volume *The Dynamics of Military Revolution, 1300–2050* contains a number of important chapters for current students of military innovation.[36] In addition to offering a comprehensive historical framework that nicely distinguishes large, epochal changes in warfare (e.g. creation of a modern nation state, French revolution, industrial revolution) with specific RMAs (e.g. steamships, combined arms tactics, submarine warfare, radar, nuclear-armed missiles, stealth, precision strike), the volume yields important insights into specific innovation periods.

Notable is Jonathan A. Bailey's chapter, "The First World War and the Birth of Modern Warfare," which outlines the advent of a new, three-dimensional approach to warfare that since the Second World War has defined much of military theory. Bailey addresses the emergence of "artillery indirect fire as the foundation of planning at the tactical, operational, and strategic levels of war" during the First World War and the subsequent "style" of warfare from which "the following ideal-type characteristics" evolved:

- It covers extended theaters and is three-dimensional.
- Time is of critical importance, in the sense of tempo—relative rate of activity—and simultaneity.
- Intelligence is the key to targeting and maneuver.
- Available hardware can engage high-value targets accurately throughout the enemy's space, either separate from or synchronized with ground contact.
- Commanders can calibrate the application of firepower to achieve specific types of effect.
- Command, control, communications (C3) systems, and styles of command that fuse the characteristics above can break the enemy's cohesion and will with catastrophic consequences.[37]

The First World War experiences with indirect fire led to profound changes in how planners and commanders conceptualized the battlefield, including an appreciation for simultaneous operations extending into the enemy's rear. In many ways we are still grappling with the implications of these developments for military organizations.

The three-dimensional "style" of warfare influenced the birth of aerial reconnaissance, which matured coordination between air and ground units, advances in precision targeting through surveys, new mapping and registration capabilities to provide unwarned barrages, and new photographic techniques (i.e. overcoming distorted images, deriving coordinates from imagery). It also pushed near-real time command and control to adjust fire, led to interception of enemy command and control communications, and a new appreciation for the relationship between fire and maneuver. Overall, warfare in the third dimension co-evolved with, and significantly reinforced the need for, C3 capabilities. During this process, as training, planning, and actual operations extended into three physical dimensions while time (the fourth dimension) was increasingly compressed, the lexicon of military thought became increasingly linked to technology underwriting C3 innovations.

Several works examine US airpower innovations. Ben Lambeth's *The Transformation of American Air Power* remains the definitive account of the US

Air Force since the Second World War.[38] Another important volume is Richard P. Hallion's *Storm Over Iraq: Air Power and the Gulf War*.[39] Kenneth Werrell's *Chasing the Silver Bullet* focuses on the evolution of precision bombing and stealth, with excellent chapters on the role played by the supporting aircraft that provide command and control, refueling, and electronic warfare capabilities.[40] Michael Russell Rip and James M. Hasik offer a technical history of precision navigation and strike weapons in their *The Precision Revolution: GPS and the Future of Aerial Warfare*.[41]

The third category informing this study includes works addressing the sources of doctrinal innovation. Frequent foci of these studies are the external influences on military doctrine during the period between the First and Second World Wars and, more generally, on internal processes causing or impeding the development of successful, innovative military doctrines. Of chief concern are influences on the emergence of particular doctrines, specifically factors leading to offensive or defensive doctrines. Although these studies are organized around military doctrine, they tend to address all of the elements of military organizations and national strategy. They also tend to study the relationship between doctrine and performance in a specific armed conflict. Notable examples are Jack Snyder's *The Ideology of the Offensive: Military Decision Making and the Disasters of 1914*, Barry R. Posen's *The Sources of Military Doctrine: France, Britain, and Germany Between the World Wars*, and Elizabeth Kier's *Imagining War: French and British Military Doctrine Between the Wars*.[42]

Relative to the first two categories, military doctrine studies are viewed as core works in the military innovation subfield. They self-consciously and deliberately explore, document, and assess military processes and outcomes to dissect doctrinal innovations that affect military effectiveness. Such studies usually engage in theory building and theory-testing, seeking to develop explanatory models about influences on military doctrine. They address the role of civilians in doctrinal innovation and the affect of organizational dynamics within military decision making processes concerning the development of doctrine.

Both Snyder and Posen found that civilian intervention in the formation of military doctrine has a positive effect when it accurately aligns military doctrine with grand strategy or other foreign policy objectives. Their agreement stems, in part, from their belief that civilian intervention to induce military change proceeds only from accurate knowledge about military affairs, or from a more developed understanding of the security environment. They also agree that organizational factors, including resources, prestige, and institutional autonomy, lead military organizations to pursue offensive doctrines.

Posen treats military doctrine as a "subcomponent of grand strategy that deals explicitly with military means" that concerns two questions: "*What* means shall be employed? and *How* shall they be employed?"[43] Using balance of power theory and organizational theory to analyze military doctrine, Posen ascribed a preponderance of influence for doctrinal innovation to civilian intervention, with civilian influence sometimes requiring military "mavericks" to be effective. Drawing primarily from the case of the British Royal Air Force's (RAF) decision to pursue air defense capabilities in the 1930s, he argues that British civilians were responsible for innovations in RAF air defense systems that later staved off and won the Battle of Britain during the Second World War.[44]

Kier disagrees with explanatory frameworks ascribing causality for doctrinal innovations to external, structural factors alone. Arguing that structural factors used by Posen, Snyder, and others are empirically indeterminate, and therefore inconclusive for theory-building, Kier turns instead to domestic politics and organizational culture for the source of doctrinal innovation. She further takes issue with organizational approaches assuming that roles and missions already performed by military organizations determine their decisions about warfighting concepts and doctrine in the future. "Deducing organizational interests from functional needs," she argues, "is too general and too imprecise."[45] Accordingly, she develops her own explanatory framework that locates answers to questions about doctrinal innovations "in domestic political battles, not foreign threats."[46] Kier maintains that civilian intervention responds to domestic, not international, politics.

In the case of the origins of the new American way of war, both internal and external factors are important. Although Kier is right in arguing for the inclusion of domestic politics, she understates the degree to which domestic political battles are conditioned and shaped by changes in strategic context. Overall, in military innovation studies—including doctrinal innovation—the idea of an innovation milieu seems a more helpful construct for discussing internal and external influences on military innovation because it does not attempt to compartmentalize or partition motivations or other influences contributing to innovation behavior. Later chapters explore the question of civilian intervention and revisit doctrinal innovation, addressing whether doctrinal or other ideational variant of innovation is really a distinct subfield of military innovation studies or merely another route to synthesizing complex organizational behavior.

A final category of works informing military innovation studies include those attuned to the diverse factors increasing or decreasing the effectiveness of military organizations within specific strategic environments. What differentiates them are their methods, scope, and case selection, which are generally more sensitive to the indeterminacy and contingency of military innovation phenomena than the other categories. This fourth category focuses quite self-consciously on military innovation as a form of social behavior—successful or unsuccessful—so that lessons, insights, or patterns might inform more contemporary policy decisions. At times sacrificing the richness of historical narrative and the objectivity of deeper case studies, these studies attempt to organize processes and behavior within specific frameworks to tease out variables and other artifacts for discussion.

Recent examples include Millet and Murray's volume on military innovation in the interwar period and Theo Farrall and Terry Terriff's previously mentioned *The Sources of Military Change: Culture, Politics, Technology*.[47] Among the earliest examples, Edward L. Katzenbach's essay "The Horse Cavalry in the Twentieth Century: A Study in Policy Response," originally published in 1958, remains an interesting study of the politics of military change.[48] Michael Armacost's *The Politics of Weapons Innovation: The Thor-Jupiter Controversy* stands as a landmark study in the role of interservice politics and service lobbying for weapons systems, documenting how both uniformed and civilian interest groups respond to international and institutional changes when arguing for weapons innovations.[49]

In this category is also the chief work on military innovation within political science, the work on which this study aims to build: Stephen Peter Rosen's *Winning the Next War: Innovation and the Modern Military*.[50] Rosen's seminal work remains the most comprehensive attempt to assess a diverse range of innovation cases (American and British) to inform post-Cold War defense modernization and transformation discussions. Building on the underlying academic question, "When and why do military organizations make major innovations in the way they fight," Rosen investigates "how the United States can and should prepare for the military problems it faces"[51] and aims to inform "Americans concerned with the possible need for military innovation."[52] Rosen analyzes twenty-one cases of successful military innovation using a threefold typology of American and British innovation cases spanning the years 1905 through 1967.

His three types of innovation include peacetime, wartime, and technological innovation:

> Peacetime military innovation may be explainable in terms of how military communities evaluate the future character of war, and how they effect change in the senior officer corps. Wartime innovation is related to the development of new measures of strategic effectiveness, effective intelligence collection, and an organization able to implement the innovation within the relatively short time of the war's duration. Technological innovation is strongly characterized by the need to develop strategies for managing uncertainty.[53]

Problems occur when applying the categories to more contemporary innovation cases. Binning military innovation into these three categories leads one into intellectual cul-de-sacs when contemplating what his cases might mean for innovation activities writ large. Generalizing from his case studies to others is therefore more difficult than necessary. In an era where information technology is both ubiquitous and a primary factor in advancing military effectiveness, and after a decade of unprecedented operational tempo, Rosen's suggestion that each of his innovation types derive from "distinct sets of intellectual and practical problems" erects false boundaries within a larger contextual milieu and proposes unhelpful analytic distinctions.

Some of the innovations discussed in Chapters 4 and 5, moreover, are likely to fall into Rosen's definition of technological innovation but also seem to support arguments from other innovation categories. This warrants restating an important analytic caution. The importance of technology, especially information technology, to peacetime and wartime innovations during the last several decades makes it increasingly difficult to retain "technological" as a distinct military innovation category. "Peacetime and wartime organizational innovation," as Rosen defines them, involve "social innovation, with changing the way men and women in organizations behave. Technological innovation is concerned with building machines."[54] This distinction seems to trivialize the social dynamics involved in technological innovation or, at the very least, conflates invention and discovery with the application of new technology to operational problems and diffusion through an organization. The latter are profoundly social.

Excellent case studies notwithstanding, his framework for analysis complicates extrapolation of findings about leadership, organizational dynamics, role of knowledge about threats, and other insightful conclusions to similar situations. For him, the emergence of new measures of military effectiveness follows from the "process of rethinking how operations lead to victory and devising new ways to measure how military capabilities relate to strategic effectiveness."[55] But his discussions of military effectiveness are primarily limited to wartime innovation. Cold War developments discussed in later chapters, not to mention 2000s defense transformation initiatives while fighting a war on terrorism, do not fit neatly into Rosen's peacetime and wartime categories.

Rosen's treatment of intelligence is also problematic for defense transformation scholars in the 2000s, for which intelligence is a key concern. Rosen concludes that "intelligence about the behavior and capabilities of the enemy has been only loosely connected to American military innovation."[56] This is certainly not true for the period studied here, nor is it the case for planners making decisions about future force structure needs.

Despite organizational and theoretical problems, however, *Winning the Next War* retains currency as a cornerstone of the evolving military innovation subfield, ostensibly because it offers a foundation amenable to revision and adaptation ten years after its 1991 publication.

On innovation for profit

The 1990s witnessed an explosion in innovation studies and innovation theories in business and management literature.[57] The business and management studies flagship journal, the *Harvard Business Review*, recently documented that innovation emerged as a top management priority only at the end of the 1990s after nearly a decade as a tertiary—at best—item on the business management agenda.[58] Thomas Kuczmarski, Arthur Middlebrooks, and Jeffrey Swaddling found in *Innovating the Corporation* that "quality" emerged as the core corporate concern in the 1980s in response to "the threat of foreign competitors offering higher quality products"; "reengineering" captured boardrooms in the early 1990s; and in the early 2000s, organizations "are beginning to publicly declare innovation as a top priority."[59]

Few military innovation scholars integrate concepts from the innovation for profit literature. They are, in many respects, not directly applicable to the study of military change. But they do provide important insights into the role of leadership, vision, and organizational culture in the successful implementation of change processes.

John D. Wolpert of IBM's Extreme Blue, a so-called "innovation incubation activity," summarizes corporate views of innovation derived from Industrial Research Institute surveys. In the "late 1980s," Wolpert relates, "most executives reported little interest in innovation" and, "in the early 1990s, innovation didn't rate among the top five corporate priorities." This changed by the end of the 1990s, when innovation emerged "at the top of the list."[60] Another shift in management theory is also being mirrored in defense modernization discussions. Early 1990s' attention on reengineering within businesses and market segments has been replaced by a focus on integration and collaboration across them. Exemplifying the change in thinking is James Champy's shift from writing on reengineering in the early 1990s to "X-engineering" (cross-engineering) in the early 2000s. "Whereas reengineering showed managers how to organize work around processes inside a company," Champy argues, "X-engineering argues that the company must now extend its processes outside" to achieve "vast improvements in operations across organizations."[61]

The shift from revolutionary, "hard right turn" management strategies to ones attuned to innovation theories and processes is arguably part of a larger drive to understand and successfully lead change in large, complex organizations operating in uncertain times. Innovation for profit theories, case studies, frameworks, and management tools are widely read and followed in the business world.

US defense transformation evolved in the 2000s as a strategic management concern for an administration that entered office with a pledge to reform the military and throughout its first term assured the nation that the war on terrorism would not forestall meaningful change. Official "strategic plans" outlined transformation objectives. Social scientists aiming to inform defense transformation cannot expect defense planners, or for that matter the larger cohort of defense transformation interlocutors, to adapt academic frameworks to their policy needs.

Political scientists interested in informing this important policy arena, for example, should not expect to achieve policy relevance unless findings are communicated in ways policy makers can readily understand. Military innovation scholars should pursue theoretical frameworks that yield conclusions decision makers can place in context with today's problems. RMA scholars succeeded in socializing their work within the defense policy community because, in the early 1990s, this community was highly receptive to, and indeed thirsted for, frameworks able to place seemingly revolutionary changes in military effectiveness in some historical perspective. Meanwhile, planners welcomed new concepts and theories into military thought as nuclear-centric thinking about defense planning lost its relevance.

Ten years later, arguably, these same planners no longer required insights into the historical dimensions of military change. Nor were they wanting for theories, concepts, or new ways of thinking about warfare. Instead, transformation planners require insights into the strategic management of military innovations pursuant to additional leaps in military effectiveness. This suggests an avenue for social scientists to further draw upon management and business school research. The earlier presentation of four military innovation categories informing this study did not include this research because it is not a traditional source for military innovation scholars.

There may be good reason for not including management theory on innovation into military innovation studies. By the late 1990s, the use of the term innovation was so widespread in business literature and management theories that Paul C. Light concluded it was "one of the most overused, underdefined [*sic*] terms in organizational life. No one seems to be sure just what the word means."[62] Strongly associated with innovation in the business world are novelty, newness, uniqueness, significant change, performance leaps, new market niches, new product creation, and the sense of more efficient resource utilization or increased value for customers.

John Kao, economist and founder of the Idea Factory, argues that the turn toward the language of innovation stems from the "imperatives of the new economy," including "speed, pushing new forms of winner-take-all competitive dynamics, introducing new business models that involve the creation of standards," and "the accelerated transformation of technology."[63] As does Light, Kao laments that innovation "is so important that the word itself is groaning under the weight of expectations placed on it. Yet as a systemic practice, it remains obscure."[64]

One sees the same befuddlement in discussions of defense transformation, a domain where more serious and systematic approaches to innovation are needed. The field, however, is rich in case studies and planning strategies offering frameworks for other scholars to utilize.

Clayton M. Christensen's *The Innovator's Dilemma* is one example of how the innovation for profit literature can inform military innovation studies. The dilemma central to his work is the historical fact that the "logical, competent decisions of management that are critical to the success of their companies are also the reasons why they lose their position of leadership" in the market.[65] "Disruptive technologies," he argues, "bring to market a very different value proposition than had been available previously. Generally, disruptive technologies underperform established products in mainstream markets."[66]

The point, also applicable to early stages of military innovation, is that major innovations that truly depart from established practices or capabilities should not be assessed against currently available capabilities. Almost by definition, a disruptive innovation (technology, organizational, or operational) requires a different approach to or set of metrics for measuring performance characteristics. This is because the means–ends relationship changes in the calculus of effectiveness. In some cases the essence of effectiveness changes.

Not all arguments and concepts from the innovation for profit literature fit with organizational dynamics and cultural attributes of military services. As Light, Peter Drucker, and others point out, important differences exist in their application in the private versus public sectors. Drucker, for example, highlights differences in degrees of change characterized as innovative or noteworthy. "In any institution other than the federal government," he argues, "the changes being trumpeted as reinventions would not even be announced, except perhaps on the bulletin board in the hallway."[67] Here the comparison is primarily based on organizational processes and other internal changes rather than on wholesale shifts in mission, customer bases, and markets. Arguably, one reason for the diverse, and sometimes contradictory, range of definitions in the business management domain is the sheer diversity of analytic interests involved. A broader range of organizational, technological, and philosophical issues exists, leading to a variety of measurements and conceptions for what constitutes significant increases in "value" for the firm. Moreover, many studies on innovation for profit are wedded to existing management philosophies or business schools, which have different analytical bents.

Although a danger exists that innovation will be overused and rendered meaningless as a term and perhaps process, military innovation studies can benefit from some of the language and ideas of those seeking innovation for profit. The diversity of leadership philosophies, change management frameworks, and corporate cultures supports, perhaps, *too* many innovation constructs. For those leading change, on the other hand, because successfully *implementing* a strategic plan is as important as the plan itself, leaders often benefit from a wide range of analytic tools and processes to bolster organizational change.

All of this begs the question of how to leverage insights for military innovation students. One area where business management studies are relevant concerns the innovation milieu itself, what Kao discusses as "the importance of physical environments that support innovation, that make innovation processes concrete, that support and generate persistence around knowledge creation processes."[68] Within the construct of an innovation milieu, moreover, business innovation case studies provide ample data on the importance of strategy and processes, aligning future needs and performance gaps with new technology, doctrine, and unit tasks. Select lessons from business management case study literature are returned to in later chapters in discussions of processes and leadership.

Although this study does not comprehensively document or exploit the innovation for profit literature, these sources remain an important sector of innovation knowledge that needs to be periodically scanned for insights into aspects of innovation common across types of organizations. They are also critical resources for leaders. One important area concerns organizational dynamics associated with diffusion and insertion processes. Here, the focus is on leading and executing innovation processes, whether they involve aligning organizations to succeed at new tasks or missions, closing critical performance gaps by implementing doctrine or other ideational changes in processes or operations, or merging organizational cultures to overcome biases and barriers impeding innovation diffusion.

Learning from the military revolution in early modern Europe

Students of military innovation can tap another source for insights into the historical dimensions of changes in warfare. The historiography of military revolutions originated in 1956 when Michael Roberts published *The Military*

Revolution, 1560–1660, which initiated an ongoing debate among historians.[69] Roberts' general field of study was early modern Sweden and his focus was Gustavus Adolphus, a focus in part derived from his biographical work on the eighteenth-century Swedish king. Examining the transformation of European warfare during the period 1560–1660, Roberts suggested four major changes to warfare in Early Modern Europe. First, there was a shift in tactics from the classic square of the Spanish *tercio* to linear formations. Thereafter, tactics based on lines of forces dominated Western warfare until well into the Industrial Revolution. Second, traditional weapons, including the lance and pike, were displaced by new weapons, including arrows and then firearms. Third, the size of armies supported by political entities increased dramatically. Finally, the overall impact of warfare and war on society grew.[70] The general argument made by Roberts was that this period witnessed a revolution in tactics, which was based on the increased size of well-drilled armies (that fought in linear formations) and which significantly increased war's impact on society.

Twenty years after Roberts published his work on the military revolution in Early Modern Europe, Geoffrey Parker expanded its foci with *The Military Revolution: Military Innovation and the Rise of the West, 1500–1800*.[71] Parker critically assessed Roberts' thesis, finding it plausible yet incomplete, and his work became the intellectual center of an expanding debate among historians. He concluded that Roberts' thesis was insensitive to changes in naval and siege warfare, overlooked military education, and ignored the codification of certain laws of war. Parker's military revolution had a wider scope, one explored through the question, "Just how did the West, initially so small and so deficient in most natural resources, become able to compensate for what it lacked through superior military and naval power?"

Much of his answer to this question revolved around the theme of action–reaction in the relationship between the offensive and defensive aspects of warfare, a relationship modulated by the introduction and rise of cannon and their effect on fortification technology. Essentially, the military revolution was linked to the development and proliferation of bastion-style fortification technology, otherwise known as the *trace italienne*.

Brian M. Downing further expanded the military revolution debate among historians with *The Military Revolution and Political Change: Origins of Democracy and Autocracy in Early Modern Europe*.[72] For Downing, "[t]he 'military revolution' or 'military modernization' refers to the process whereby small, decentralized, self-equipped feudal hosts were replaced by increasingly large, centrally financed and supplied armies that equipped themselves with ever more sophisticated and expensive weaponry. The expense of the military revolution led to financial and constitutional strain, as parsimonious and parochial estates refused to approve the requisite taxes."[73] Downing's military revolution is political, rather than military, in its consequences and definition, and he argued more forcefully for thinking of weapons and armaments as only a small part of a military revolution. For him, a military revolution involved more than a single combat arm or technology issue, an argument he made by exploring the complex politico-military changes transpiring in the sixteenth and seventeenth centuries. His version of the military revolution focused on the social and political conditions wrought by war. They led some constitutionalist societies in Europe to evolve into "military-bureaucratic absolutist" states while others developed into liberal democracies. Generally, states that did not have to deal with high domestic pressures to mobilize and support war avoided the need to develop highly centralized, absolutist governments.

Clifford Rogers is another influential voice in the historical debate on military revolutions. He argues that a "focus on the centuries after 1500 obscures the importance of the period in which the most dramatic, most truly revolutionary changes in European military affairs took place: the period, roughly, of the Hundred Years' War (1337–1453)."[74] During the Hundred Years' War, he contends, European war was revolutionized twice, first by an infantry revolution (which matured in the middle of the fourteenth century) and second by an artillery one (which occurred in the first third of the fifteenth century). Rogers states that these two revolutions were followed in later centuries by two other military revolutions in Early Modern Europe. A fortification technology revolution, which was the centerpiece of Parker's military revolution, was followed by a revolution in administration, which was a focus of Roberts' military revolution.

Rogers' conclusion is interesting in light of how military theorists approached the task of defining an "American RMA" in the early 1990s. After stating the obvious—that the "concept of 'revolution' in history is a flexible one"—he restated his observation that not one but a *series* of military revolutions occurred between 1300 and 1800. All four of the above mentioned revolutions were "synergistically combined to create the Western military superiority of the eighteenth century."[75] The identification of four military revolutions in close historical proximity led Rogers to ponder whether the period was actually one of long evolution rather than four distinct revolutions. Suspecting that the overall theory of military revolutions might be impeding our understanding of the natural order of things, he borrowed the concept of "punctuated equilibrium evolution" from biology.

Punctuated equilibrium, he argued, might be applicable to the history and theory of warfare. Under such an approach, evolution is characterized by short periods of rapid alteration followed by long periods of "near stasis" in which only slow, incremental changes occur. Indeed, it appears that the processes of innovation and transformation, which antecede RMAs, adhere more to the model of punctuated equilibrium than to the idealized revolutionary construct implied in much RMA literature.

Noteworthy in the historical debate over the true boundaries and features of the military revolution in early modern Europe is the lack of quibbling over definitions and the general detachment from the question, "What's in a name?" Each of these military historians, only a few of which are discussed above, produced a well-researched, painstakingly argued, and historically accurate argument that bounds a particular research question within a certain analytical context. The methodology employed by each follows a traditional historical method, although Downing's is more a comparative political history.

Studies of the military revolution in early modern Europe do not offer insights into organizational behavior or political processes directly transferable to today's transformation discussions. These works inform students of the American RMA and provide insights into the utility of academic debate over the unfolding of history. Each of the above works clearly articulates the image or definition of the military revolution under consideration, with caveats about the merits and demerits of the analysis. Each also displays deep interest in social, political, organizational, economic, and other "non-technological" aspects of warfare. Finally, these works demonstrate sensitivity to the nature of fundamental change from one period of history to the next.

A more noteworthy aspect of these studies concerns Rogers' reflection on the approach to military revolutions. Parenthetically, it appears that he and other historians were intellectually *compelled* to engage in the historical military revolution debate established within the discipline by luminaries such as Roberts and Parker. Indeed, Rogers steps back from his argument to consider the greater issues of what he is studying and attempts to give an alternate perspective for the theory of military revolutions, a perspective drawn from biology's theory of punctuated equilibrium. In doing so he all but admits that he *has* to call his focus of analysis a "military revolution" to be accepted within the discipline even though his study does not align with others. He is forced to fit his argument within the discourse on military revolutions despite the fact that, presumably, he finds the approach lacking because it treats military revolutions as discrete, temporally if not causally.

This is a syndrome students of US defense transformation need to avoid, one that appears to have infected post-Cold War defense policy discussions. By turning to a military innovation framework, which is amenable to unlimited adaptation to facilitate explorations of innovation variables, students of US defense transformation can avoid the intellectual *cul de sac* Rogers decried.

Chapter conclusion

Historical awareness is, for sure, crucial to navigating arguments for and against military change. Murray points out that "no example in history" exists "where military organizations have successfully jumped into the future without a compass from the past to suggest how they might best incorporate technology into a larger framework."[76] Just as business leaders carefully select cases and theories applicable to their needs, so too should defense transformation scholars focus on applicable historical cases.

One motivation for this study is the paucity of historical study on the late Cold War period. The preponderance of innovation cases referenced in military innovation works during the 1990s, for example, addressed interwar military innovations. Fiftieth anniversary reflections on the epochal events of the Second World War, including blockbuster movies and best-selling histories, likely increased the attractiveness of interwar studies among defense planners responsible for defense transformation. What is interesting in retrospect, and indeed a motivation for this study, is the lack of analysis on the period that gave rise to capabilities labeled revolutionary in the aftermath of the 1990–1991 Gulf War.

Barry Watts and Murray ascribe the "motivation" underlying interwar studies to a "*hypothesis* that" the US was "in the early stages of a period in which advances in precision weaponry, sensing and surveillance, computational and information-processing capabilities, and related systems will trigger substantial changes" in military capabilities "as profound and far reaching as the combined-systems 'revolutions' of the interwar period."[77] Chief among interwar references used to frame the American experience was the German combined-arms armored capabilities popularly known as blitzkrieg or "lightening war." As Murray concluded elsewhere, the German's 1940 "breakthrough on the Meuse" in northern France "and its explosive exploitation…was so crushing, so convincing, that it has served as the shining exemplar of *the* revolution in military affairs of the mid-twentieth century."[78]

Parallels emerged in the way the German and American RMAs were defined, the former lending images and concepts to the latter. Both involved relative advances in command and control of distributed forces. Blitzkrieg was described as a combined-systems RMA consisting of radio communications, tanks, tactical air cover, doctrine, and operational practices; the American RMA became known as a "system-of-systems" revolution defined primarily by advanced surveillance, information warfare, stealth, long-range precision strike, and joint warfighting doctrine.

Former Secretary of Defense William Perry (1994 through 1997), an early proponent of a system-of-systems approach and father of some of the military capabilities observers labeled an RMA in the 1990s, describes the underlying philosophy of system-of-systems thinking as "links in the chain of effectiveness." If "any one of these had been removed, the overall effectiveness of the chain would have been significantly diminished."[79] Crucial to keeping these links was the coevolution of doctrine and technology.

The US experience differed from the German interwar in one important respect. It included the emergence of what Israeli military historian Shimon Naveh terms an "operational cognition" in the last decades of the Cold War. Where German military leaders failed to develop an operational cognition during the interwar period, American commanders did in the 1970s and 1980s.

The Germans certainly achieved tactical excellence. Operational excellence, including theater-level planning and an ability to coordinate operational maneuver, was lacking. Barry Watts and Williamson Murray concluded that Germany was able to demonstrate revolutionary capabilities because it "evolved sound concepts for mobile, combined-arm warfare and had trained their army to execute those concepts."[80] But at the operational and strategic level, Naveh faults Germany for "deep operational ignorance" in the conduct of the Second World War and, during major campaigns, adhering to a "strategic framework" that "completely disregarded the existence of depth, space, fighting resources and operational trends."[81]

Above the tactical level, "Blitzkrieg not only lacked operational coherence but…[in] its actual formation dictated relinquishing a systemic approach to military conduct."[82] The German Army lost its ability to think operationally in the late 1930s largely because Hitler severely weakened the German staff school and crushed the effectiveness of a core group of military thinkers that appeared unsympathetic to Nazi ideology. The training of military officers thereafter "centered on the levels of the brigade and the division and only rarely touched on problems related to the operating of corps."[83] Matthew Cooper's more thorough study, *The German Army, 1933–1945: Its Political and Military Failure* reinforces key elements of this argument, which he calls the "myth of the Blitzkrieg."[84]

During the 1980s, as part of a larger reawakening of American military thought, the operational level of war became a key focus of study and an important consideration in defense planning. In the planning domain, this involved thinking of mission packages in which all the required aspects of an operational capability were developed and fielded together, with subelements integrated into an enterprise. Discursive and organizational parallels to this operational cognition included a range of images and activities associated with different forms or archetypes of integration. Enterprises, networks, network-centric operational concepts, common operating pictures, joint organizations and doctrine, and other terms were evoked to describe idealized behaviors and conditions all concerned with synergistic, emergent capabilities of the Services acting in concert.

For Naveh, the United States developed the first true systems approach to military planning in the West because it includes a cognitive orientation, a schema, conditioned to think about military operations in a systems-theoretic fashion. Strategic objectives, campaign plans, and tactics are linked.[85] A larger integration theme coevolved with the emerging information technology sector, leading to network centric warfare and other concepts. This was the core of the offset strategy, discussed in Chapter 4.

3 American military strategy from the Second World War through Vietnam

This chapter reviews key aspects of US military strategy from the end of the Second World War through the rise of flexible response strategy in the late 1960s. The sections below introduce strategic policy issues and provide background information helpful to understanding military innovations discussed in subsequent chapters.

For the majority of the period discussed in this chapter, national military strategy and military doctrine were primarily concerned with nuclear strategy and nuclear war planning. This is no longer the case. A new generation of American policy makers, strategists, and military analysts has matured within a post–Cold War national security environment that is no longer dominated by discussions of nuclear war planning. This new generation is intellectually free of the paradox of nuclear deterrence theory and the deceptive simplicity of a defense strategy organized around the strategic triad, the Cold War deterrent force consisting of nuclear-armed bombers, silo-based intercontinental ballistic missiles, and submarine-launched ballistic missiles.

For decades, nonnuclear or conventional warfighting forces were first and foremost considered adjunct capabilities on the margins of nuclear strategy. From the perspective of Cold War American military strategy, it is remarkable that, in the early 2000s, a new line of tactical nuclear weapons is being considered to reinforce existing nonnuclear global strategic strike forces in their mission of destroying hardened underground facilities.

To fully grasp the innovations that occurred in the 1970s and 1980s, we must first understand the broad strokes of the historical conditions that created a moment ripe for military innovation and defense strategy reforms. We cannot characterize what many in the 1990s termed an American revolution in military affairs without knowing what the revolution overturned. Those proclaiming the arrival of a new American way of war in the 2000s, furthermore, have neither defined what was left behind nor provided a sense of what has been retained in strategic culture, military thought, or defense planning from the height of the Cold War period.

During the period explored in this chapter, nuclear deterrence evolved into a grand strategy. When deterrence credibility issues arose in the 1970s, nonnuclear dimensions of military thought and defense planning received renewed attention after years of relative neglect. The United States subsequently pursued an ambitious military innovation agenda that laid the foundations for American military primacy in the 2000s, the story we will turn to in Chapters 4 and 5.

On "Carrying a Twig": postwar defense planning

In the post-Second World War period, US national security planners grappled with vexing domestic issues while redefining the nation's role in global politics. Domestic issues included transitioning the economy off wartime price controls, managing post-Second World War demobilization, and sustaining military preparedness as defense spending declined.[1] On the international security agenda, pressing concerns included rebuilding Europe and containing Russia.

Then Secretary of State James Byrnes characterized the essential national security planning challenge with a clever turn of Theodore Roosevelt's statement, "Uncle Sam should speak softly and carry a big stick." Alluding to the possibility of conflict with Russia and wary of Russian conventional military advantages, Brynes lamented that reduced defense spending and demobilization left him in the awkward position of speaking loudly but carrying a twig.[2]

"Some of the people who yelled the loudest for me to adopt a firm attitude toward Russia," Brynes later observed, "yelled even louder for the rapid demobilization of the Army."[3] Personnel strength declined by some ten and a half million between June 30, 1945 and June 30, 1947, as depicted in Table 3.1[4]

Continued military engagement abroad was unpopular at home, as were proposals to maintain high levels of defense spending. Servicemen would accept nothing less than a speedy return home and immediate transition to civilian life; riots resulted when delays occurred. Efforts at re-conversion to a peacetime economy continued as wartime price controls were lifted, government rationing was relaxed, labor policies restricting wages and job security were eased, and regulations on the production of goods requiring rubber, iron, or other war-related material were removed.

Meanwhile, Joseph Stalin transitioned Soviet wartime economic controls into a command economy charged with reconstituting and modernizing Russian military forces. Demobilization was pursued with an eye toward rapid reconstitution of military power. Some eight million men left Russia's armed forces from 1945 to 1948, leaving sizeable stores of weapons and equipment for the three million remaining in uniform. Of these, some 80 percent remained in the ground forces

Table 3.1 Armed forces strength

	June 30, 1945	*June 30, 1946*	*June 30, 1947*
Army (less Army Air Corps)	5,984,114	1,434,175	683,837
Army Air Force	2,282,259	455,515	305,827
Navy	3,377,840	951,930	477,384
Marine Corps	476,709	155,592	92,222
	12,120,922	2,997,212	1,559,270

that would soon be occupying Eastern Europe. Stalin allocated sizeable industrial resources toward defense preparedness. Research and development initiatives produced new weapons, including the T-54 and T-55 battle tanks and innovative battlefield missile systems. Entire sectors of industry were focused on advanced electronics and other military support systems. The American monopoly on nuclear weapons and the need to maintain control over the Soviet sphere of influence were powerful incentives to sustain Soviet military power.

The emerging security environment was placed in perspective by George Kennan's 8,000-word "Long Telegram." This penetrating, remarkable analysis of the Soviet threat was received in Washington on February 22, 1946. Widely circulated in Washington, his analysis of Soviet foreign policy objectives gave historical depth and structural coherence to what was a muddle of arguments about likely Russian behavior. Kennan's memo provided insight into the likely motives underlying seemingly erratic Soviet foreign policy. It also outlined why an understanding of the social, economic, and historic realities underlying Soviet strategic culture was critical to shaping successful American policy.[5]

Kennan suggested a policy course in his telegram. Arguing that conflict with the Soviet Union was not inevitable, Kennan reasoned that, while the Soviets would expand their influence wherever possible, their actual behavior could be moderated. Steady opposition to expansionism through political and economic policies, not military threats, would be sufficient. He further argued that America should not expect Soviet actions to conform to Western visions of behavior or Western preferences for open diplomatic relationships grounded in the moderation of power. To counter the Soviet's power politics, Kennan prescribed patient and firm containment consisting of political and economic measures to offset or blunt expansionist behavior.[6]

The idea of adopting a patient containment policy was sharply criticized during the 1946 congressional campaign. Widespread criticism reinforced existing, downward trends in public opinion. Truman's approval rating declined from 80 to nearly 30 percent while fiscal and political conservatives from across the political spectrum criticized his policies. Republicans won majorities in both the House and Senate in the November 1946 Congressional elections. Within a year, containment would be redefined in more military terms and thenceforward provide the organizing principle of Cold War American policy toward the Soviet Union.[7]

Recognizing that the incoming Congress would move quickly to revamp defense planning and challenge his management of defense issues, Truman commissioned studies of defense requirements and readiness level, including assessment of munitions supplies and logistics requirements. Despite criticism of foreign policy and security issues, however, the domestic political environment did not support an increase in defense spending to increase readiness or to develop new military systems. The Joint Chiefs of Staff developed joint strategic concepts to refine mobilization and procurement, and defense planning slowly matured. But the planning process neglected the most important area in defense strategy, atomic weapons.

Henry Kissinger, doyen of American national security analysts and archetypal foreign policy adviser, cogently captures one of the themes in the evolution of

American military strategy during the Cold War. It is a central theme in this study. Describing the early Cold War defense planning environment, he lamented the "gap between military and national power" that developed because the atomic bomb was added to the American arsenal "without integrating its implications into our thinking."[8] John Lewis Gaddis adds that President Truman "and his advisers were as uncertain about what they could actually *do* with nuclear weapons when they left office in 1953 as they had been in 1949."[9]

What was the overall effect? Military historian Russell F. Weigley, the scholar who first popularized the notion of an American way of war in a book entitled *The American Way of War: A History of United States Military Strategy and Policy*, captured the fallacy inherent in early Cold War strategic thinking by naming the relevant chapter, "American Strategy in Perplexity."[10] Arguably, the rigor many see as lacking in early Cold War strategic discourse surfaced in the late 1970s and 1980s with the rise of a more balanced view of the role to be played by both nuclear and nonnuclear capabilities.

A number of factors impeded atomic war planning, including the extreme secrecy that continued to surround all matters related to the nascent atomic arsenal and the basics of atomic weapons. Truman himself remained unaware of the exact size of the atomic arsenal until early April 1947.[11] The military did not even have access to bomb components, and knowledge of how to actually assemble and deploy the weapons was not passed from the now demobilized wartime cadre of scientists to the post-war military.

Limited bomb-production capabilities led to what historians later called the "scarcity" problem. Few weapons could have been readied for operational deployment. Only two weapons existed in 1945, thirteen in 1947, and fifty in 1948—and the bombs were actually stored disassembled. Readying them took some forty men nearly two days.[12] Assembly personnel that did remain in service were not easy to recall to duty.

Delivery options were also limited. Only a few dozen B-29 bombers were available and they remained so vulnerable to enemy anti-aircraft defenses that some doubted the weapons would ever get close to intended targets. "Because of limited capability and inadequate intelligence," Rosenberg concluded, "bomber crews could only hope to penetrate to their targets under cover of darkness and bad weather" to "locate precise aim points"—a task made more difficult over "often snow-covered targets."[13]

Strategists struggled to define a set of conceptual and operational approaches to atomic-era warfare in 1947, leading to the first atomic targeting plan.[14] This initial operational plan for the American atomic arsenal reflected the Second World War strategic bombing experience. Dubbed "city busting" attacks by some, the objective was attacking and destroying an enemy's industrial base, which typically meant industrial zones adjacent to dense urban areas. Planners believed that disrupting the industrial base and population centers would critically reduce the support to, and capacity of, an adversary's armies.

Others questioned the soundness of an atomic attack on industrial centers. Arsenal limitations and scarce delivery options meant that only a small portion of Soviet industrial capabilities could be destroyed. If Soviet war-making capabilities were not crippled, experience with bombing population centers during the Second World War suggested that the limited attacks might actually embolden an adversary's will to fight.[15] Atomic, later nuclear, targeting theory would dominate defense strategy and military thought for much of the Cold War.

Changes in the international security environment

American defense planners soon faced additional challenges. Among the most important from a strategic planning perspective involved the strength of America's staunchest ally, Great Britain.

Struggling economically, electricity rationing limited some British households to mere hours of power per day. Food scarcity emerged as a crisis across Europe, driving up food prices. Unemployment reached six million, double that of the 1930s depression. Burdened by domestic economic ills, London reduced its security commitments abroad. On February 21, 1947, Britain notified the US State Department that their economic and military aid to Greece and Turkey would end within six months. India was on a path to independence in 1948. The British mandate in Palestine was transferred to the United Nations.

London urged the United States to assume a greater role in global politics. The United States historically had approached new commitments abroad reluctantly. Typically, this is attributed to an American preference for isolationism in foreign policy. This is not necessarily accurate. A more appropriate term might be unilateralism, or the steadfast belief that engagement in the world should avoid cumbersome entanglements so that America was free to act in its own interests, on American terms. Here, it is important to understand that the United States was a reluctant hegemon, slow to accept post-war primacy despite the fact that America emerged from the war the first undisputed economic and military superpower.

The Central Intelligence Agency (CIA) was created in 1947 to provide American policy makers with insights into global security issues and to assess the capabilities of potential adversaries. It concluded that the "poverty and underprivileged position of the population" of post-colonial areas, along with "the existence of leftist elements within them," rendered "them peculiarly susceptible to Soviet penetration."[16] Debates ensued about how committed the United States should be in European economic recovery and whether to leave occupation troops in Europe.

Truman sought funding from Congress to prevent the potential loss of Greece to communist insurgents and the expansion of Soviet influence into Turkey. His March 12, 1947 speech requesting aid was dubbed the Truman Doctrine. Addressing a Harvard commencement a month later, Secretary of State George Marshall, who replaced Byrnes, unveiled a plan for European recovery that the press labeled the Marshall Plan. It aimed to underwrite Europe's economic recovery and postwar reconstruction to prevent the conditions CIA analysts considered ripe for the spread of communism. Willingness to counter Soviet expansionism increased in response to the Soviet-orchestrated 1948 coup in Czechoslovakia and the increasing popularity of communist movements within some local European political parties.

Pentagon planners recognized that defense spending was not the most important strategic priority: European reconstruction was more important for the long-term defense of Western Europe. As Melvyn P. Leffler documents in *A Preponderance of Power: National Security, the Truman Administration, and the Cold War*, while few believed Moscow would engage in military aggression in the near future, many believed economic problems threatened "the long-termed balance of power" and required immediate attention.[17]

A March 1948 report by the Joint Ad Hoc Committee, which included CIA, State, Army, Navy, and Air Force intelligence officers, assessed that Moscow would not pursue a direct military confrontation or action during 1948. Instead, the Committee predicted the Soviets would seek greater influence in Europe and other strategically important regions by exploiting and, if warranted, fomenting, political-economic crises.[18] This assessment was negated by Moscow's June 1948 attempt to use military forces to deny the United States, France, and Britain access to Berlin.

A massive airlift was conducted to prevent starvation and to signal continued American commitment to Berlin's status as an open city. Postwar agreements partitioned Germany into four occupation zones, one for each of the Big Three and a fourth for France. Berlin was designated an "open-city" within the Soviet-controlled zone. Using air corridors guaranteed by postwar agreements, supplies were airlifted to the besieged city for three-hundred and twenty-four consecutive days with a peak daily resupply rate of some 13,000 tons. It was a remarkable show of airpower.

The United States avoided a direct military confrontation during the eleven months of the Berlin Blockade. A skirmish with ground troops might spark a more serious military standoff or even a war, neither of which the United States was prepared for militarily. When analysts on both the sides of the Atlantic questioned America's military preparedness to respond to a more serious crisis, it appeared that they were also questioning the credibility of the American commitments to defend Western Europe.

Given the relatively poor state of conventional preparedness, the crucial issue became the deterrent value of the American atomic arsenal. Although no atomic bombs were actually assembled and deployed to Europe, a fact Soviet intelligence agents reported to the Kremlin during the airlift, Truman's decision to deploy atomic-capable bombers to British bases was believed by some to have demonstrated American resolve. Regardless of the actual effect the bomber deployment had on Soviet leaders, American strategists did not have many options for offsetting Soviet conventional superiority in Europe. If East–West relations deteriorated into outright war, conventional wisdom held that Western Europe would be overrun unless America's atomic arsenal was formally linked to its defense. Once Soviet forces penetrated into Western occupation zones, atomic weapons were deemed the only recourse to compel Moscow to withdraw. Atomic war planning thereafter figured more prominently in defense strategy.

This spurred a 1948 "extended deterrence" pledge committing America to the atomic defense of Western Europe. The pledge was formalized with the April 1948 creation of the North Atlantic Treaty Organization (NATO). Five months later, Truman signed a National Security Council policy directive on the use of atomic weapons for NATO's defense. National Security Council Memorandum 30 (NSC-30) directed that the US armed forces "must be ready to utilize promptly and effectively all appropriate means available, including atomic weapons."[19] This impelled additional consideration of when, how, and where atomic weapons might be used, including targeting scenarios. For the next decade, NSC-30 served as the principle document outlining atomic (later nuclear) weapons employment policies.[20]

Another document, NSC-20/4, outlined a set of strategic objectives to guide war planning, stating that the goal was to simply reduce or eliminate Soviet "control inside and outside the Soviet Union," not annihilating the Soviet regime altogether.[21] NSC-30 and NSC-20/4 changed the course of defense planning and military strategy by calling for a formal atomic war planning process and, indirectly, by addressing the atomic readiness problem by setting the expectation that the weapons would be available for operational use. Neither document, however, was effective in resolving larger issues of how, when, where, and why "the bomb" would be used, including how to actually overcome the above-mentioned limitations in the arsenal itself. These issues would be left unresolved until the Eisenhower administration (1953–1961).

Subsequently, military innovation during the Cold War generally derived from perceived changes in deterrence stability and associated threats to the European "extended deterrence" pledge. NATO defense planning requirements and a 1949 Joint Chiefs of Staff study had already pointed toward new targeting priorities for the US Strategic Air Command. The Harmon Report, named for its chair, Air Force Lieutenant General H.R. Harmon, concluded that early use of atomic weapons remained crucial to damaging Soviet war-making capabilities. As planners had already recognized, too few atomic weapons existed to destroy all targets. It restated concerns that a limited bombing campaign might merely bolster Soviet will to continue fighting.

The Soviet Union tested an atomic bomb in June 1949, ending the American atomic monopoly and further complicating defense planning. The CIA concluded that the advent of a Soviet atomic arsenal would increase the USSR's "military and political capabilities" and the "possibility of war"; overall, the CIA assessed that "the security of the United States is in increasing jeopardy."[22] Communists took control of China in December, increasing fears that the world was slipping toward a new world war between democratic and communist blocks.

European security remained the overarching concern. The Strategic Air Command was assigned the task of developing atomic plans for the defense of Western Europe, a process that shifted the focus of planning from industrial centers and strategic war-making capabilities to the actual *retarding* of Soviet military forces.[23] President Truman authorized the expansion of the atomic arsenal, approved a more aggressive weapons development and testing program, and, on January 31, 1950, approved development of a much more powerful thermonuclear hydrogen bomb (H-bomb).

On the same day Truman approved the H-bomb he directed the Secretaries of State and Defense to reexamine strategic plans. The response came in February 1949 from the State Department's Policy Planning Staff, led by Paul Nitze, in the form of NSC-68.

NSC-68 was the first comprehensive national security document of the Cold War to link international security conditions to specific defense spending and force structure proposals. Among its policy recommendations was a massive US arms buildup. NSC-68 served as an ideological and conceptual justification for the H-bomb decision by reinforcing a pessimistic image of Soviet intentions and predicting the continued deterioration of US–Soviet relations. It urged adoption of a containment strategy backed by increased defense spending.

Another noteworthy aspect of NSC-68 was the outright rejection of a declared no first-use policy concerning nuclear weapons. "In our present situation of relative unpreparedness in conventional weapons," NSC-68 stated, the Soviets would consider an American no-first use policy "an admission of great weakness"; America's allies would interpret it as "a clear indication that we intended to abandon them" in a conflict.[24] The emerging logic of nuclear deterrence held that Moscow could be prevented from expanding westward only through an absolutely credible threat to use nuclear weapons in response to any Soviet attack.

NSC-68's recommendations to increase defense spending would likely not have been acted on had North Korea not crossed the thirty-eighth parallel into South Korea in June 1950. Some argued that North Korean invasion "saved" NATO by reversing the direction of defense preparedness.

Increased support for defense spending was in part related to widespread, but misplaced, beliefs that communist North Korea was being directed, or at least aided, by Moscow. Fears of Soviet expansionism had been fueled by a 1949 report, NSC-48-1. It concluded that the Soviet Union sought to dominate all of Asia. Losing Korea might start a "domino effect" leading to other regimes falling to communist aggression, subversion, or subterfuge.

America's military intervention in Korea led to a reappraisal of NSC-68's argument for a tougher stance toward Moscow and increased defense spending to offset Soviet numerical advantages in Europe. One outcome was an important shift in the essence of containment strategy. Whereas Kennan envisioned patient and firm containment through political and economic means, the post-Korean War view emphasized a greater role for military power and less room for patience.

The new look

President Dwight D. Eisenhower assumed office in 1953, just as the Cold War was emerging as a full-blown international contest with implications for US national security far beyond post-Second World War European occupation zones. During his two terms in office, the United States took its first steps toward military involvement in Indochina, the Middle East emerged as an arena of conflict after Egypt aligned with Moscow, Cuba became the first Soviet client state in the Western Hemisphere, the United States intervened in the Congo, and Pacific Rim security issues received additional scrutiny.

Despite hopeful signs that relations with the USSR might improve following Joseph Stalin's death on March 5, 1953, East–West relations continued to decline. The Soviet Union's brutal repression of a workers' riot in Berlin that May reminded Western analysts of Nazism and Fascism. An unpredictable, hard-liner named Nikita S. Krushchev assumed leadership of the Soviet Communist Party in 1955. The same year the Soviet Union signed a defense and security cooperation agreement with its Easter European clients in Warsaw, Poland (the Warsaw Pact) to oppose NATO. In November 1956, some 200,000 Soviet troops and over 5,000 Soviet tanks crushed a Hungarian uprising, killing 20,000 protestors opposing Soviet occupation.

In light of global security challenges, Eisenhower moved quickly to revamp defense planning. Eisenhower brought important knowledge and experience about military and strategic affairs to the White House. In addition to his Second World War leadership experience, he served as Supreme Allied Commander of NATO forces in Europe. During his time commanding NATO, he oversaw the development of the first plans to defeat numerically superior Soviet forces with nuclear weapons. He understood the emerging role nuclear weapons might play in European security and their utility—and their limits—as battlefield weapons.

The mid-1950s witnessed important changes in American nuclear strategy. Fearing Soviet interference during the Korean War, and forced to plan for the threat of a Soviet atomic attack in Europe, the priority for nuclear targeting shifted from general war-making capacity to direct targeting of Soviet nuclear facilities. This included delivery vehicles, which at the time meant airbases. The mission of "blunting" Soviet atomic capabilities became the highest priority. The second priority would be "retarding" Soviet forces engaged in an attack on NATO, with the lowest priority given to the more traditional attacks to disrupt or destroy Soviet war-making capabilities (e.g. fuel, power, and atomic industries).[25]

Like Truman, Eisenhower faced political opposition to increased defense spending. Critics charged that the administration was doing too little to counter the negative effects increased defense spending had on the consumer economy during the Korean War. Massive armament and military modernization efforts during the Korean War, some argued, were impeding economic growth by constraining labor mobility and keeping interest rates from rising.

Eisenhower proposed a "New Look" defense policy to align fiscal realities with security challenges. He aimed to simultaneously balance the budget, reduce defense spending, cut manpower, and improve the nuclear arsenal. Regarding nuclear weapons, Eisenhower believed that the implied threat to use nuclear weapons during the Korean War impelled the North Koreans to sign an armistice and end hostilities in July 1953.

The New Look shifted the burden of conventional defense spending from US troops stationed abroad to regional allies. A series of alliances and security treaties expanded to some 150 nations, many outside of Europe. Of note were Eisenhower's October 23, 1954 pledge of American assistance to South Vietnam and the formation of the Southeast Asian Treaty Organization (SEATO) to prevent communism from spreading to Vietnam, Cambodia, and Laos. By the end of the decade there were nearly one-thousand US military personnel in Vietnam, a number that peaked at some 544,000 troops in 1968.

The intent of regional security arrangements was straightforward: incorporate indigenous forces into regional defense schemes for planning purposes to reduce, at least in theory, the burden on American military forces. The agreements also called for the creation of local forces to fight communist expansion. Increased reliance on indigenous forces and decreased reliance on American troops would decrease the overall planning and funding requirements for conventional forces. More resources could therefore be allocated to nuclear force modernization, which Eisenhower believed was critical for deterring the Soviet Union. Subsequently, spending on strategic nuclear forces rose steadily while the overall growth of defense spending declined.

With scant resources for nonnuclear weapons systems, little effort was given to modernizing conventional forces. Second World War stockpiles remained the primary source of equipment and supply well into the 1950s. US Army priorities in the early 1950s remained occupation and civil defense duties, tasks that did not embolden the development of new ground forces operational doctrine or pose challenges to stimulate innovative new conventional weapons capabilities. What innovations did transpire involved miniaturization and other technologies to provide small, lower-yield tactical nuclear weapons for nuclear artillery and landmines.

Underscoring the New Look were important changes in nuclear doctrine. Eisenhower likened nuclear weapons employment decisions to conventional ones. "Where these things are used on strictly military targets and for strictly military purposes," he opined, "I see no reason why they shouldn't be used exactly as you would use a bullet or anything else."[26] NSC 162/2, "Basic National Security Policy," was signed in late October 1953. It maintained that, if conflict occurred with either Russia or China, US nuclear weapons would "be available for use as other munitions."[27]

This view of nuclear weapons shaped the first true US nuclear strategy declaration—massive retaliation—linked to a fully coordinated, comprehensive set of force structure decisions. US forces adopted a massive retaliation planning assumption in 1953 and the Administration publicly declared the policy in 1954. Air power remained the favored delivery mechanism.

Massive retaliation left open the possibility that the US might meet any aggression or threats to its interests with a devastating nuclear attack. Capitalizing on the inherent ambiguity in a policy based on nuclear retaliation to *any* attack, the policy intended to create such uncertainty among Russian leaders about the consequences of a war that they would demur from any act of military aggression.

Massive retaliation was adopted by NATO's Military Committee (MC) on December 17, 1954, with the approval of MC 48. While nuclear weapons had always been part of NATO military planning, MC 48 recognized a larger role for them by emphasizing the use of nuclear weapons for the defense of Europe. "What was special" about this shift, historian and strategist Marc Trachtenberg notes, was the planning "assumption that there was one, and only one, way in which the Soviets could be prevented from overrunning Europe in the event of war, and that was through the very rapid and massive use of nuclear weapons, both tactically and strategically."[28] As a defense strategy, because it emphasized an immediate, massive nuclear response, some detractors referred to this as the "spasm" war plan.

To bolster the implied threat of an American nuclear defense of any regional security partner, American policy makers continued to reject a "no first use" declaration. In the face of Soviet aggression, the United States reserved the right to use nuclear weapons first, even if it meant a nuclear strike before a Soviet attack (if an attack on NATO appeared imminent).

Trachtenberg believes that MC 48, considered in tandem with the application of no first use, was in essence a preemptive strategy, although not stated in such terms due to the sensitivity of a formally declared preemptive nuclear doctrine. In language repeated some fifty years later by George W. Bush, Eisenhower argued in a 1954 National Security Council meeting that the United States "must not allow the enemy to strike the first blow" and later sought military capabilities to initiate a preemptive attack.[29] "Victory or defeat," Eisenhower wrote, "could hang upon minutes and seconds used decisively at top speed or tragically wasted in indecision."[30] A 1981 Office of the Secretary of Defense history of this period characterized the Eisenhower defense strategy as one in which all available resources were focused on developing a "rapid (indeed preemptive) and massive response to an imminent attack."[31]

Trachtenberg's summary of MC 48 as a military strategy provides insights into themes we will revisit in later chapters:

> What this meant was that if America was to get in the first nuclear blow—viewed as vital to the survival of the western world, and thus a prime goal of the strategy—she would in all likelihood have to launch her attack *before* the enemy had actually struck. Hence the great emphasis on speed, something that would have made little sense if America and NATO as a whole had opted for a simple retaliatory strategy in the normal sense of the term.[32]

Decision speed, the doctrinal ideal of massive retaliation through overwhelming shock, and the potential for preemption were all reinforced in military strategy when Eisenhower approved the development of the Thor and Atlas missiles. Thor was America's first intermediate-range ballistic missile (IRBM). Atlas, the first US intercontinental ballistic missile (ICBM), became operational in 1959.

Soviet missile developments had been a concern to American strategists through-out the decade. A 1954 intelligence estimate, for example, reported "a large and active research and development program"; it concluded, however, that the actual threat from a surprise attack could not be further assessed without "firm current intelligence" on what the USSR was developing or had deployed.[33] This changed with the 1957 launch of the Sputnik satellite, which demonstrated the advanced state of Soviet rockets and delivered a blow to American pride. Sputnik reinforced concerns that Soviet ICBMs would soon be deployed, providing Moscow with an additional deterrent against the United States implementing extended deterrence policies if the USSR moved against NATO. Soviet rocketry advances raised the issue of a "missile gap," which John F. Kennedy later exploited in his successful 1960 presidential campaign against Eisenhower's Vice President, Richard Nixon.

Memories of Pearl Harbor and Korea combined with the development of missiles able to strike anywhere in the globe within thirty minutes to fuel fears of surprise attacks. Preparing for a surprise attack became a central facet of American defense planning. Several programs emerged to increase US intelligence collection behind the Iron Curtain and to illuminate Soviet capabilities as well as intentions. Understanding Soviet military developments and assessing their impact on deterrence became an overriding national security priority, rivaled only by the issue of protecting US retaliatory forces from being destroyed in a so-called "bolt-from-the-blue" attack.[34] Such intelligence reporting informed the decision to aggressively pursue an American strategic missile force. Programs such as the U-2 reconnaissance plane and space-borne intelligence collection from Corona spy satellites became central to defense planning and to the building of targeting capabilities, leading to American preeminence in technical intelligence gathering.

The coming of Soviet missiles led the Eisenhower administration to adapt its deterrence posture from massive retaliation to what became known as "graduated deterrence." As then Secretary of State John Foster Dulles opined in 1957, this led to "less reliance upon deterrence of vast retaliatory power."[35] Eisenhower himself came to question the logic of a massive retaliation doctrine as early as 1955, concluding that it provided " 'no defense against the losses we incur through the enemy's political and military nibbling. So long as he abstains from doing anything that he believes would provide the free world to an open declaration of major war, he need not fear' " the deterrent power of America's nuclear arsenal.[36]

In addition to changing the parameters of strategic warning, when both sides had missiles the decision making time available for preparing retaliatory strikes decreased. Missiles therefore required different types and levels of strategic planning and targeting preparation. Planners increasingly questioned whether it still made sense to threaten massive retaliation, simply because any US retaliation with weapons based in Europe could bring a Soviet strike on the continental United States. The assumptions underlying massive retaliation were revised, leading to a new label for the overall strategy of deterring any attack with the threat of nuclear retaliation, "mutually assured destruction" or MAD.

At the end of the Eisenhower administration, in August 1960, then Secretary of Defense Thomas Gates created a Joint Strategic Targeting Planning Staff to maintain data on all targets warranting attack in a US nuclear strike (the National Strategic Target List, NSTL) and to prepare target assignments for all US nuclear forces (the Single Integrated Operational Plan, SIOP). The US nuclear arsenal had now reached some 18,000 weapons, with some 90 percent of them under direct military control, an outcome of Eisenhower's treatment of nuclear weapons as any other item in the military stockpile.[37]

Gates wanted the NSTL and SIOP completed by December for the incoming administration. The complete list included over 2,000 targets in the Soviet Union and China, ranging from ICBM bases to command and control centers to at least 131 urban centers. Targeting, planners subsequently argued, "should involve a series of 'sequential options', consisting of such targets sets as 'central strategic systems, theater threats, and countervalue targets' [economic and industrial targets, including cities]."[38] The groundwork was thus set for a change in targeting doctrine and additional shifts in military strategy.

Flexible response and the emerging framework for innovation

Lawrence Freedman argues that the underlying strategic foundation of the New Look was a signal to Soviet leadership that the "West would not reply in kind to an Eastern invasion but raise the stakes of war" with nuclear weapons. "Thereafter," he continues in his classic *The Evolution of Nuclear Strategy*, "Western strategy would depend on convincing the Soviet leaders that it had the nerve to do this. This problem would become progressively more difficult as the Soviet capabilities to fight at the new level increased."[39] Indeed, it became more difficult still after Soviet capabilities increased in both the nuclear and conventional domains.

As Soviet capabilities in both areas increased through the late 1960s and 1970s, Soviet leaders seemingly had two trumps to the West's sole nuclear threat, their own nuclear forces and the massive Red Army.[40] This situation, as Chapter 5 discusses, eventually spurred US defense planners to respond with a large conventional modernization program in the late 1970s and early 1980s, a program wholly concerned with stabilizing deterrence stability in central Europe. Its subsequent adaptation, however, would later underwrite the development of regional intervention capabilities and an offensive, rapid-dominance approach to warfare.

NATO's collective deterrent against a Soviet attack in Europe rested, or so it seemed, on NATO's ability to communicate a *credible* deterrent threat, one that conveyed the Allies' commitment to use nuclear weapons.

Military forces deter in several ways. Their very existence presents obstacles to an aggressor, in political or military terms (or both), increasing the uncertainty inherent in cost–benefit projections figuring in decisions to attack. This was, of course, the notion underscoring the presence of US forces in Berlin. Another way military forces contribute to deterrence is the implied threat of retaliation in kind or through escalation. The key is holding at risk something that an adversary values; its potential destruction makes any attack too costly to pursue. The classic example is threatening to retaliate against an adversary's population centers.

Military forces specifically arrayed in a defensive posture deter by raising the possibility that an attack might fail or, at the very least, raise the cost of succeeding. Because the defense might succeed, the attacker must commit, and therefore risk, more forces in the opening attack. This is what US and NATO military planners sought through advanced conventional forces to offset Soviet numerical and in some areas qualitative advantages in Europe.

During the 1960s, the quest for deterrence stability led to renewed interest in conventional capabilities, making the calculus of deterrence more complex. Partly this reflected the emergence of new paths to restore deterrence credibility by offsetting Soviet capabilities. One outcome was a flexible response strategy, which originated in the 1950s but did not mature as strategic defense policy until the 1960s when President John F. Kennedy questioned the soundness of MAD.

Kennedy was among the critics of Eisenhower's massive retaliation policy in the 1950s. Others included Deputy Chief of Staff of the Army Lieutenant General James M. Gavin and Chief of Staff of the Army General Maxwell D. Taylor. Both retired in the last years of the Eisenhower administration and wrote critical books that greatly influenced a national debate on defense policy. Gavin's *War and Peace in the Space Age* argued for greater Army efforts in the fields of missiles, the development of air mobility, and the pursuit of tactical nuclear weapons.[41] Taylor's *The Uncertain Trumpet* was more influential. He introduced the term "flexible response" into the defense lexicon. The underlying concepts originated in Britain during the 1950s within the larger framework of "graduated deterrence" and could also be found in Eisenhower administration discussions of nuclear strategy. Taylor was recalled to active duty by Kennedy as a special military advisor and then as Chairman of the Joint Chiefs of Staff.

Rejection of the Eisenhower administration's declared strategy of massive retaliation policy spilled over to nuclear target planning. The existing SIOP, known as SIOP-62, was anything but flexible when the Kennedy administration assumed office. As soon as nuclear war was initiated with the Soviet Union, SIOP-62 called for launching all US strategic weapons at predesignated targets. No weapons were withheld, leading to the "Spasm-War" euphemism. Soviet and Chinese casualty estimates approached half a billion.[42] The fatalism of the plan, its inherent rejection of other escalatory options, and the revelation that "whatever happened, some portion of the admittedly inferior Soviet long-range force would survive to strike America," led the administration to reject SIOP-62 and build on the work Gates initiated at the end of the Eisenhower administration.[43]

The Cuban Missile Crisis, another Berlin crisis, Third World proxy wars, and an overall deterioration in US–Soviet relations reinforced the search for a new nuclear strategy, one that would "match the potential range of challenges with a correspondingly broad range of options."[44] Kennedy thought that the weaknesses of US and NATO conventional forces limited his options for dealing with the Soviets, forcing him to rely on somewhat incredulous threats to use nuclear weapons.

One element of the new strategy was greater appreciation of, and the ability to respond to, nonnuclear conflicts, which among other developments spurred increased funding for special operations (unconventional) forces. Kennedy subsequently doubled the number of ships in the Navy, increased tactical Air Force squadrons, and created five new Army divisions. Some of these divisions saw their first action in Vietnam; US combat forces departed for what would become America's longest war in May 1965. As later chapters discuss, Vietnam derailed plans to fund conventional modernization and to give flexible response meaningful teeth as a national strategy.

Massive retaliation had been conceived at a time when the US enjoyed a monopoly in nuclear weapon delivery systems. By the early 1960s, the Soviet Union had amassed a huge arsenal, including ICBMs. The nature of the deterrence game shifted in 1963, which Mark Trachtenberg terms "a watershed year" because the Soviet Union achieved rough parity in nuclear weapons and the Cold War became "a different type of conflict" altogether.[45] Now, for example, strategists ruminated about the balance of forces required for *intrawar* deterrence through escalatory levels of retaliation, providing time for the adversary to reconsider the efficacy of the conflict. Such thinking focused strategists on the integration of nuclear and nonnuclear capabilities required to provide sufficient flexibility for intrawar deterrence.

When Defense Secretary Robert S. McNamara formally declared a "flexible response" strategy in February 1962, the underlying focus was on refining the nuclear dimensions of national deterrence strategy to reduce reliance on nuclear weapons.[46] Defense planners revamped requirements for greater *internal* flexibility across the US military. As a planning factor, as stated above, this renewed calls for increased spending on conventional forces. And as a warfighting strategy, it emphasized the development of more capable conventional forces to fight other conventional armies. Nuclear weapons remained available under flexible response for any scenario, including first use. Officially adopted by the NATO in 1967, flexible response struggled to gain credibility.

Because NATO did not increase its conventional capabilities commensurate with the Soviet threat, the flexible response strategy left few real options other than using nuclear weapons unless the US committed additional resources. That is, men and material to bolster the Alliance's nonnuclear posture. But declining US defense spending seemed to leave few options other than theater nuclear capabilities. A political complication was added when the Soviet Union publicly declared a no-first doctrine. Few believed that the Kremlin would actually cast aside its nuclear planning; most recognized the declaration as a political statement targeted at the European masses that increasingly saw themselves caught between hair-trigger nuclear arsenals. As political pressure increased within NATO to find alternatives to early nuclear use, NATO defense planners sought new conventional alternatives to blunt Soviet aggression. *Not* increasing conventional capabilities meant the continuing Soviet buildup would increase reliance on early nuclear use at a time when political pressure against the very existence of tactical nuclear weapons was building.

Political scientist Samuel Huntington captured the essence of the evolution of flexible response, a story revisited in later chapters. Figure 3.1 adapts Huntington's model to suggest a role for preemption. It views flexible response from the perspective of a defense planner contemplating "four possible means of deterring

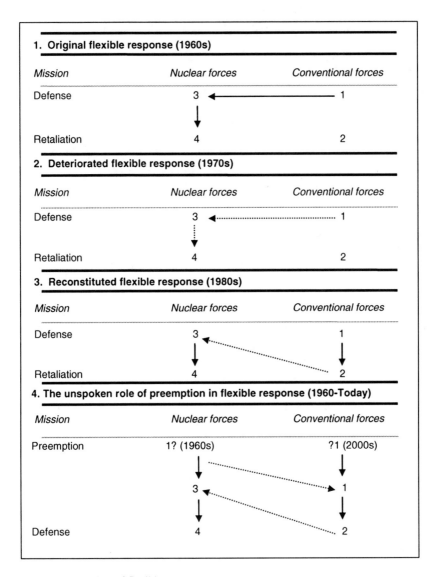

Figure 3.1 Evolution of flexible response.

Soviet aggression": conventional defense, conventional retaliation, nuclear defense, and nuclear retaliation.[47] The arrows in the figure indicate the strength of the connections between stages of the defense and retaliation options.

Parts 1–3 of Figure 3.1 are the generally accepted view of how American military strategy evolved during the Cold War after the concept of flexible response became doctrine in the 1960s. As we discuss below, however, declassified records suggest that, from the early days of flexible response, preemption figured prominently in military strategy and targeting discussions. Underlying the evolution of American military thought, then, is the argument for preemption as a necessary policy option.

At the top of the figure is the original variant for flexible response. It held that, if Soviet forces attacked, NATO would mount a conventional defense but rely primarily on nuclear defense and retaliation capabilities to halt and roll back Soviet forces. Analysts considered the American nuclear deterrent to be quite robust in the early 1960s until the Soviet Union approached nuclear parity and possessed a survivable second strike force. In the 1970s, the credibility of flexible response was called into question largely because conventional force readiness plummeted.

Defense and retaliation are key concepts. A credible deterrence regime, according to the general understanding of flexible response, must be able to both defend against attack and to retaliate in a way that threatens additional harm on the attacker. Deterring Moscow from attacking required not only the ability to mount a defense but also the threat of retaliation. This could be retaliation in kind or escalation.

Some believed that flexible response might even promote the development of "rules" (or expected behavior) concerning nuclear use. At the time national security planners welcomed "rules" or at least a common understanding of how deterrence threats could be better adjusted during a crisis. The Cuban Missile crisis, for example, had taught a lesson: negotiating pauses could be critical to the diffusion of tensions. Declaring that nuclear weapons would be used in any situation in which NATO was attacked correlated to using them in all situations, a strategy that made for a limited ability to learn how to diffuse tensions in rapidly changing, uncertain situations.

The real focus of flexible response as a nuclear warfighting strategy was a more flexible SIOP, one that would allow "controlled response and negotiating pauses in the event of thermonuclear war."[48] Flexibility before and *within* nuclear warfighting scenarios meant more elongated escalation processes before all-out nuclear war; in other words, more options and entry points for nuclear weapons in overall deterrence policy. This is a key concept: flexible retaliation preserves deterrence at all levels and types of conflicts, an important characteristic for national security planners facing uncertainty. For planners in 1960 it added a range of conventional deterrence options to what was then a relatively short list of the pre-nuclear ones.

In reality, flexible response did little to diversify nuclear-centric national security strategy and military doctrine. Tactical airpower doctrine, for example, remained focused on nuclear missions. Training manuals for the F-100 tactical fighter-bomber went so far as to instruct pilots that "nuclear training will in every instance take precedence over nonnuclear" training and pilot qualification, leading to diminished readiness for close air support missions.[49]

Proponents of conventional modernization contended that NATO decision processes for conventional defense and retaliation were nowhere near as cumbersome or prolonged as were those for battlefield nuclear weapons. Additionally, ambiguities in the NATO document adopting flexible response, MC

14/3, did nothing to resolve intra-Alliance disagreements about the aims of flexible response. European interpretations generally favored readiness to prosecute only a limited conventional defense that quickly escalated to nuclear use. In this view, conventional forces were meant to deter conventional attacks for limited gains and delay all-out attacks long enough for nuclear weapons release decisions to be made. The European view derived from a belief that greater emphasis on conventional capabilities would be perceived as diminishing confidence in NATO's willingness to use nukes. US planners preferred a more robust conventional phase.

The mismatch between flexible response strategy and planning guidance for controlling escalation during a conflict received presidential attention in the late 1960s and early 1970s. Incoming Nixon administration Secretary of Defense Clark Clifford announced a program to develop additional, more complete targeting plans with data processing centers to develop response options for future SIOPs. This involved: "(1) providing pre-planned options for the National Command Authority (NCA) for additional selected responses against military and industrial targets (e.g., strategic strikes for support of NATO); and (2) providing the procedures, data processing equipment, and computer programs for planning new, selective responses on a timely basis during the crises."[50]

Limited nuclear options, countervailing strategies, and the so-called Schlesinger Doctrine—all proposed during the 1970s—were in part based on more flexible targeting theory developed in the early 1960s but never codified in planning documents. Credibility issues continued to drive innovations in nuclear doctrine during the 1970s and 1980s. Some believed that the adoption of a flexible response strategy undermined the extended deterrence pledge at the core of NATO's defenses. After the Soviets attained nuclear parity, many feared that a limited Soviet attack might be successful and that Moscow would pursue what analysts dubbed salami tactics. That is, Soviet forces would advance only far enough into NATO territory to achieve a foothold needed for negotiations.

As later chapters will discuss, when military planners turned to conventional weapons innovations in the 1970s and 1980s to reclaim the deterrence stability achieved during the earlier periods, they proceeded along the same basic arguments outlined in the flexible response doctrine. In fact, many of the innovations anteceding the new American way of war were partly conceived to bolster the conventional options inherent in a revived flexible response strategy.

The Soviet nuclear RMA: on the strategic and operational threat

On the other side of the Iron Curtain, the Soviets adapted their doctrine and force posture to counter flexible response strategy. This created new strategic and operational challenges for the United States and NATO. The US Office of Technology Assessment concluded that, after Soviet Premier Nikita Khrushchev's 1964 ouster and the 1967 adoption of flexible response by NATO, the Kremlin also "began to consider the possibility of a war remaining conventional" and subsequently reinforced their conventional forces.[51]

In doing so, Soviet strategists added conventional elements to an existing body of military theory describing a decades-long nuclear revolution in military affairs (RMA). Soviet military analyst Joseph Douglass concludes that the nuclear revolution thesis was "an accepted precept" among Soviet military theorists by the early 1960s.[52] The nuclear RMA encompassed revolutionary changes derived from "fundamental, qualitative changes in the means of armed conflict, of methods of combat actions, in the organization of troops," and in military "training and education."[53]

When it was later applied to US forces, few American military theorists realized that the term RMA had special meaning in Soviet military thought and was associated with a well-developed set of concepts, arguments, and theories. It was a reference to similarities between advanced conventional capabilities and the decades-long discourse on multifaceted changes in strategy and warfare wrought by nuclear weapons.

Initial thinking on the nuclear RMA was described in the so-called "Special Collection" of documents passed to the West by British-directed spy Oleg Penkovsky.[54] Compiled in 1958, a year before the creation of the Soviet Strategic Rocket Forces, the papers discussed the impact nuclear missiles were having on Soviet military strategy. Khrushchev, for example, believed that nuclear weapons "made huge infantry and tank armies redundant" and dramatically reduced Soviet ground forces.[55] Khrushchev went as far as abolishing the "chief of the ground forces" position in September 1964, a month before his own ouster.

The Soviet nuclear arsenal matured thanks to years of funding increases for the strategic rocket forces. After several years of inattention to conventional forces, Soviet theorists and planners in the mid-1960s returned to core themes of mechanization, maneuver, deep battle, and combined arms attacks. It was almost as if the Khrushchev period hadn't occurred. Kimberly Martin Zisk concludes that, although debate continued on the impact of nuclear weapons on operations after Stalin's 1953 death, and despite Khrushchev, Soviet defense planners tended "to view nuclear weapons as support forces for the ground troops, not independent strike forces that obviated the need for armies."[56]

Nonetheless, no consensus emerged among Soviet analysts on the operational role for ground forces on the nuclear battlefield during this period. Many assumed that nuclear weapons would obliterate enemy forces, obviating the need for large armies. This position paralleled Khrushchev's. Other analysts considered conventional forces useful only for "mopping up" after nuclear exchanges and for restoring order through martial law. Another vision for ground forces involved enormous tank formations rolling through areas after an initial "preparation" with nuclear and even chemical attacks. Robert A. Doughty describes the early effects of nuclear thinking:

> the Soviets shifted from a primary focus on continental land warfare to a focus on global nuclear warfare. Military leaders believed that the revolution in military affairs compelled complete revisions in strategy, tactics, and force structure. Red Army commanders modified their thinking about the conduct

of ground operations in the nuclear age and emphasized dispersion, mobility, high operating tempos, and multiple attacks on broad axes.[57]

In his *Race to the Swift: Thoughts of Twenty-First Century Warfare*, Richard Simpkin details the evolution of Soviet deep operations theory and the effect nuclear weapons had on Soviet approaches to warfighting.[58] The initial phase of Soviet thinking about the nuclear RMA extended to roughly 1967, the year NATO adopted a flexible response strategy. During this period, planners assumed any conflict would lead to nuclear escalation.

Michael MccGwire sees the 1967–1968 period as "a watershed in Soviet defense policy."[59] The essential change was a switch from believing any world war would lead to nuclear "strikes against the homelands of the two superpowers" to believing "that a world war might be waged with conventional weapons" without nuclear escalation.[60]

One benchmark during this period is Operation Dnieper, a 1967 training exercise named for the river across which massed armored forces rehearsed river-bridging missions. In his *The Soviet High Command, 1967–1989*, Dale Heerspring contrasts Dnieper with the 1965 "October Storm" and the 1966 "Vlatva" exercises. "Whereas both previous exercises had included a conventional phase, each had quickly escalated to the use of nuclear weapons. This time, the exercise was entirely conventional."[61]

What factors contributed to this change? In their study of Soviet conventional warfare, Douglass and Hoeber suggest four: revisions to American and NATO strategy; awareness that the Soviet military needed additional training to fight a nonnuclear war; an understanding that nonnuclear operations would have to be designed at both the unit and subunit levels; and newfound appreciation for the limited utility of initiating a combat operation with a massive nuclear attack.[62] All of this reaffirmed a central tenet in Soviet strategic culture, that "ground forces are what, in the end, are used to implement the Soviet offensive strategy."[63] In 1967, the Kremlin also revived the position of chief of the ground forces.

Noteworthy was the degree to which Soviet planners adopted a vision of future conflict centered around integrated nuclear and conventional operations on the same battlefield. The nuclear RMA construct conditioned commanders to accept the integration of nonnuclear forces, operationally and doctrinally, into military affairs, leading to the "rapid development and mass introduction of nuclear weapons, missiles, and radio electronic means among the troops as well as the significant improvement of other types of armament and combat equipment."[64] Doctrinal and operational integration of "new weapons in all categories of the armed forces" impelled "radical changes in the methods of conducting warfare" and a "review of the established principles of the art of war."[65]

While US planners held similar views, Soviet planning seemed more sensitive to the exigencies of planning, equipping, and training for conventional operations. Only in the late 1970s would the US focus greater attention and resources on an "integrated" battlefield.

Soviet reactions to NATO's adoption of flexible response doctrine, specifically the doctrine's approach to managing escalation from a conventional defense phase to increasingly more severe nuclear strikes, brought another period of competition between NATO and the Warsaw Pact. As defense analyst Steven Canby concluded in a 1973 report for the Defense Advanced Research Project Agency, flexible response "tends to invite an enemy strategy of lightening invasion, against which NATO is least prepared to defend, and subsequent negotiation in order to fragment an alliance under stress."[66] Indeed, this is how Soviet strategy evolved, a development addressed in Chapters 4 and 5.

The threat of a lightening armored strike into NATO territory, even for a limited gain from which to improve negotiations, increased concern among NATO defense intellectuals that Soviet forces would strike before nuclear weapons could be employed. Some analysts predicted it would take as long as a week to finalize a decision to use tactical nuclear weapons to defend NATO. Conventional retaliation options, made credible by more capable and ready forces, would allow NATO field commanders to simultaneously defend NATO's front lines and attack advancing Soviet forces.

A CIA memorandum on Soviet Defense Policy from 1962 through 1972 concluded that, "the Soviet view of war in Europe had undergone a significant change" and now reflected a belief "that the initial period of a war with NATO could be fought without the use of nuclear weapons."[67] The CIA also concluded that "Soviet acceptance of a possible nonnuclear phase of hostilities [led] to some changes in force structure."[68]

Soviet force structure changes included adding a motorized rifle division to each tank army and "five new tactical air defense systems, five artillery systems, three new tracked combat vehicles, and improved tactical engineering and logistics systems."[69] BMP1 armored infantry fighting vehicles, the first mass-produced modern armored vehicle since the Second World War, were introduced to speed infantry forces into battle alongside new T-72 main battle tanks. BMP1 armament included Sagger anti-tank missiles and a new anti-tank gun. Anti-tank weapons became ubiquitous. "Division artillery," analysts discovered, "increased by about 50 percent"—and they were self-propelled. Mechanized forces became much more capable in terms of bringing fire support to bear on a fluid battlefield.[70]

Soviet military planners also pursued the "fullest possible use of the airspace."[71] Airborne (air-deliverable) versions of the BMP and other armored vehicles were fielded. The ZSU-23 air defense system, with its four 23-mm cannons, was the most sophisticated in the world. Russian divisions advanced within a "bubble" of protective air defense systems, including ZSU-23s for defending against low altitude aircraft and a suite of surface to air missiles (SAMs) for higher ones. This required interlocking air defense radar, warning, and fire control systems.

It was during this period Zisk argues, when US defense planners were focused on Vietnam, "that the Soviet General Staff began to implement its infamous 'conventional option'. Fully developed by the late 1970s, it held that Soviet conventional weapons would be used to destroy NATO theater nuclear weapons before they could be used, and thus secure victory ... without the use of nuclear weapons."[72] Because the poor state of NATO conventional forces would require escalation to nuclear weapons, "it might indeed make sense for the Soviets to try to make them as useless as possible as soon as possible, and to try to accomplish as much as possible before they came into play."[73]

American military strategy and the legacy of Vietnam

Vietnam was a crystallizing event of the Cold War era and continues to weigh heavily on the American collective consciousness. President Lyndon Johnson famously decried it as a "bitch of a war." George McGovern characterized it as America's "second civil war" and rightly predicted that his generation would be fighting it "for the rest of our lives."[74] Some thirty years after US troops left Vietnam, the most contentious and widely covered issues during the 2004 presidential campaign concerned John Kerry's war record as a swift boat commander in Vietnam and George W. Bush's duty status in the Texas Air National Guard during the war.

The Vietnam experience was a tumultuous one for the American military establishment and remains a turning point in the history of American military strategy. Lacking a conventional defense and retaliation alternative, awakening to the post-Vietnam imperative for radical reform, and succumbing to political pressure at home and abroad to reduce reliance on nuclear deterrence, American national security planners undertook a series of initiatives in the shadow of Vietnam that coalesced into a significant increase in US military effectiveness in the 2000s.

Throughout the 1950s and 1960s, American military strategists placed high priority on nuclear missions in the development of doctrine. Chapters 4 and 5 will examine the consequences for defense planning and deterrence stability after the Soviet Union achieved parity in nuclear weapons and greatly increased their conventional forces, two developments that occurred while America was preoccupied with the Vietnam War. Perhaps the more important post-Vietnam change in military strategy was the final recognition that conventional forces were in need of significant reform. "By worrying about strategic weapons," former CIA Director William Colby argued in 1977, "we will indeed be fighting the wrong war" by failing to address conventional, economic, and political components required "to meet the threat."[75]

Some defense analysts feared that, in the shadow of Vietnam, nuclear deterrence strategy would continue to impede the development of nonnuclear military strategy and that the United States would grow reluctant to engage militarily abroad. Albert Wohlstetter was among the most outspoken advocates of revamping military strategy and defense planning. In the decade following American withdrawal from Vietnam, he was among a growing number of military reformers that believed new technology and new operational practices could yield more decisive nonnuclear military capabilities. These capabilities, moreover, could then be used to pursue American interests abroad and to restore the credibility of American military power.

Bounding the period discussed in this chapter are two distinct periods of social-political tension in American military history. At the beginning of the Cold War, servicemen rioted in Europe against perceived delays in demobilization. Recognizing that the troops would remain abroad for occupation duty, a debate occurred among policy makers grappling with America's post-Second World War global responsibilities and status as the world's most powerful nation. At the end of the period discussed in this chapter, a growing peace movement placed pressure on political leaders to end the American military presence in Vietnam.

In Europe, racial riots on US military bases, rampant drug use among American serviceman, and general tensions with Allies all served to undermine confidence in NATO's ability to mount a credible conventional defense against a Soviet attack. For much of the 1970s, American units in Europe remained anything but combat effective. A strategy of flexible response was not supported by defense spending, doctrine, or operational preparedness. Scant force modernization occurred during the decade of involvement in Southeast Asia. American equipment staged in Europe to defend NATO was poorly maintained. Allies openly questioned the US commitment to a serious conventional defense.

Post-Vietnam US foreign policy behavior did not resolve Allied concerns about American capabilities to defend against a Soviet attack. During the Nixon and Ford administrations, events pointed to US retrenchment or at least reluctance to involve US military forces in regional conflicts. When Saigon fell in 1975, Ford was unable to secure military and economic aid for the struggling South Vietnamese. Later that year, Congress passed the Clark amendment limiting military and economic aid to pro-Western forces fighting Soviet-backed communists in Angola. Military interventions abroad to contain Soviet regional expansion was politically untenable for much of the decade.

Although strategic and operational necessity increasingly called for conventional force innovation in the decade following American withdrawal from Vietnam, nuclear force issues continued to dominate defense planning. Political support for conventional force modernization did not materialize until the late 1970s; only in the 1980s did revolutionary conventional warfighting capabilities begin to enter the force.

As had been the case during the 1950s, nuclear weapons modernization was pursued throughout the 1970s and 1980s to bolster deterrence. National security strategy remained pinned to the strategic nuclear triad of strategic bombers on ready alert, extremely quiet submarines carrying ballistic missiles, and land-based ICBMS housed in protective underground missile silos. Conventional forces in Europe were a "tripwire," a symbolic representation of America's commitment to the Allies and an attempt to scribe a "line in the sand" that Moscow would have to cross. Although Soviet forces could easily overrun these forces, doing so would trip, or trigger, nuclear war. As became clear in the late 1970s and early 1980s, however, conventional forces were needed inside and outside of the European theater to offset Soviet conventional military power.

Advances in Soviet general-purpose forces in the 1970s renewed debate over the offensive–defensive balance in Europe and refocused attention on the credibility of nuclear deterrence. Some discussion occurred about the adequacy of European defense spending, discussions that were eclipsed by far-ranging debates about force modernization and doctrinal innovation. Ultimately, America's European allies were unable, for economic and political reasons, to increase their defense budgets to modernize their conventional forces.

During the debates about how to offset Soviet conventional capabilities in Europe, proponents of advanced conventional warfighting forces found their ideas thrust onto the center stage. Alternatives to a nuclear-centric defense of Europe resonated with the post-Vietnam generation of military leaders seeking to restore an offensive spirit to American military thought and to provide the tools needed for victory in any future regional conflict. In this climate, a cadre of military officers set about transforming training, doctrine, and other aspects of warfighting on the margins of the scramble for new strategic systems able to preserve the deterrent value of the venerable triad.

Before leaving the immediate post-Vietnam era it is instructive to consider one soldier's view of how Vietnam influenced future military leaders. Retired Army Lieutenant General Pat Hughes served as a platoon leader with the 9th Infantry Division in the Mekong Delta (1969) and as advisor to a provisional reconnaissance unit in the Phoenix Program (1971–1972). His last job in uniform was as Director of the Defense Intelligence Agency (DIA). In the early 2000s, he returned to government service as a civilian to lead information integration and intelligence activities at the Department of Homeland Security.

Many who know him well affectionately refer to him as "Yoda" because of his deep philosophical understanding of the contemporary American military, how it has struggled in the past, and how it is best applied to overcoming America's foes abroad as well as recovering its purpose as a positive force in American society.

Vietnam, Hughes contends, forced a generation of American military leaders to contemplate the utter "finality" of combat at a time when they were rethinking the Second World War generation's unabashed commitment to military service. For him, military thought during much of the early Cold War was wrapped in a highly abstract cloak of nuclear targeting; it never possessed the clarity of the Second World War or even the Korean conflict. Calculations of blast effects, radiation, and overpressure were the work of civilian technicians adding scientific detail to the work of civilian theorists—the political scientists and economists shaping deterrence theory. A defense discourse centered on abstract nuclear warfare issues left tacticians and planners unprepared for unconventional war in Vietnam.

Militarily, politically, operationally, Vietnam was unlike the two world wars of liberation or the inaptly named Korean "conflict" that was more than a police action for those who fought it. In Vietnam, especially at the tactical level, where the battlefield climate of chaos and destruction imparted a vivid, deeply personal appreciation for the non-linearity of even meticulously planned operations, respect for the finality of war led post-Vietnam Army officers to profess that they would never allow another Vietnam to occur.

Quite simply, Hughes recalls, they had to go to war to win. That is, the military had to be given a clear set of objectives, given the resources required to accomplish them, and provided strong political backing to implement plans. It is difficult to relate the historic importance of this sentiment to those that did not live through the experience. Nor is it easy to convey the angst felt by a generation of military leaders concerned that they might *not* go into battle with the political backing and resources to win. It is important to recognize how the views of Hughes and other officers responsible for strategic planning, technology development, and doctrine were driven by their Vietnam experience.[76]

Chapter conclusion

For Martin van Creveld, the evolution of military strategy during the period discussed above resulted in the "splintering" of strategy into "nuclear strategy, conventional strategy, grand strategy, theater strategy, economic strategy, and other types too numerous to mention."[77] Van Creveld laments that the discussion of "strategy" was itself rendered virtually meaningless. By the end of the Cold War, the term "became one of the buzzwords of the age, meaning the methodical use of resources to achieve any goal, from selling consumer goods to winning a woman. In the process, it lost most of its connections with the conduct of large-scale war."[78]

Lawrence Freedman made a similar argument in an October 1955 article entitled "Strategy Hits a Dead End." Freedman found that "the position we have reached is one where stability depends on something that is more the antithesis of strategy than its apotheosis—on threats that will get out of hand, that we might act irrationally, that possibly through inadvertence we could set in motion a process that in its development and conclusion would be beyond human control and comprehension."Decrying the very idea of a nuclear "strategy" in the traditional sense of the term, Freedman would later conclude his seminal work on the history of nuclear strategy by simply stating, "C'est magnifique, mais ce n'est pas stratégie."[79]

Few captured the larger shift in military thought associated with the dawn of the atomic age as elegantly or persuasively as did Bernard Brodie in his 1946 treatise, *The Absolute Weapon: Atomic Power and World Order*.[80] Just months after Hiroshima, Brodie argued, "Thus far the chief purpose of our military establishment has been to win wars. From now on its chief purpose must be to avert them. It can have almost no other useful purpose."[81] So it seemed with the evolution of nuclear strategy and its dominance in American defense planning. But in a world in which nonnuclear forces were needed to fight and win small wars, there were indeed other useful purposes for military forces.

A theme from the above sections concerns the *business* or organizational processes associated with the implementation of various nuclear strategies in the sizing and fielding of military forces. These are essentially decisions about what mix of military capabilities, nuclear and nonnuclear, would be pursued to address the range of possible military missions or conflicts. As the above discussion implies, during much of the Cold War nuclear war planning and nuclear strike capabilities were highly privileged within the defense planning process to the detriment of conventional force planning, doctrine, and training.

Attention did eventually swing toward nonnuclear defense preparedness. As the next chapter argues, international and domestic factors impelled the Carter administration to reverse key elements in its national security strategy. Grand strategy was refashioned to place greater emphasis on nonnuclear defense issues as the United States expanded its military presence in the Middle East.

4 Military innovation in the shadow of Vietnam
The offset strategy

The story of Cold War American defense policy is often told through the history of specific ideas or concepts. Terms like Iron Curtain, containment, the Marshall Plan, the Truman Doctrine, massive retaliation, flexible response, détente, and others simply yet meaningfully capture the essence of a complex issue, strategic relationship, or policy position. In a period in which strategic discourse and military strategy was dominated by such terms, it is curious that the relatively straightforward "offset strategy" concept discussed in this chapter has been all but ignored by students of Cold War American defense strategy.

In the late 1970s, Secretary of Defense Harold Brown and Undersecretary of Defense for Research and Engineering William Perry conceived of an "offset strategy" to address strategic and operational challenges posed by Soviet forces. The technologically advanced weapon systems associated with the offset strategy, Perry recalls, "were conceived and developed during the 1970s" as the corner-stone of America's "response to the then-perceived threat of an armored assault by the Warsaw Pact forces in central Europe."[1] As Soviet forces became more capable, defense planners feared that a rapid armored assault would penetrate deep enough to destroy NATO's tactical nuclear forces, thereby removing "a step in the NATO escalation ladder" and prevent NATO from mounting a nuclear defense entirely.[2]

Reflecting in the aftermath of the 1991 Gulf War about the legacy of the 1970s, Perry concluded that during "the 1970s U.S. defense officials saw the opportunity to exploit the new developments in microelectronics and computers to great advantage in military applications." The United States "conceived, developed, tested, produced and deployed the systems embodying the new technologies" and "developed the tactics for using the new systems, and conducted extensive training with them, mostly under simulated field conditions."[3] Perry argued that the offset strategy, which "sought to use technology as an equalizer or 'force multiplier," was in fact "pursued consistently by five administrations during the 1970s and 1980s."[4]

Certainly there was no claim then or even in this study that the term or idea of an "offset strategy" is by itself historically noteworthy. Military history is replete with examples where states adopt or adapt technological, organizational, and operational innovations to compensate for a weakness in one's own forces in a way that offsets an adversary's strengths. What was new in the late 1970s and early 1980s? Among the most important changes was *how* defense planners and military leaders adapted their visions for future fighting forces to include information technology. As information and computer technology enabled new capabilities, and as new definitions of military effectiveness emerged, information and decision capabilities dominated visions for the future of American military power. What made the offset strategy noteworthy was how new capabilities enabled better decisions and more effective actions across space and through time. To offset numerical superiority, American forces would leverage the knowledge advantages accrued by integrated surveillance, targeting, and command networks to increase the speed and precision of attacks.

The calculus of American military readiness shifted in the 1970s as the United States assumed military commitments outside of Europe. Defense strategists on both sides of the Atlantic scrambled to understand how Soviet-American competition in the second and third world might influence nuclear deterrence in the European theater. Increasingly, it seemed that the ability to deter Soviet aggression in strategically important regions like the Persian Gulf was linked to deterrence stability in Europe. As the decade unfolded, national security planners pursued alternate paths for stabilizing deterrence that tied European issues with important new regional challenges. These paths led defense planners and military strategists to nonnuclear options, including the origins of rapid deployment, rapid dominance, and information superiority.

Defense planners and military officers responsible for preparing and leading troops to defend NATO's front lines rejected the notion that American troops were a mere tripwire, a force that Soviet armored formations would overrun before being destroyed by tactical nuclear weapons. Military officers did not accept the mission of defending and most likely dying in place as NATO political leaders debated the use of nuclear weapons; to do so ran against the post-Vietnam desire to rebuild the armed forces. The emerging generation of military leaders sought to transform the armed forces technologically, doctrinally, and operationally, to prepare to fight and win the opening battles of any future conflict, and to instill an offensive spirit within their respective service cultures.

The post-Vietnam security environment

The highlight of President Richard M. Nixon's first week in office was the signing of the January 1972 Paris Peace Settlement ending America's military presence in Vietnam. An important benchmark in the history of American foreign policy and defense planning, the Settlement marked the beginning of the thirty-year transformation in American military effectiveness that culminated with the 2003 invasion of Iraq.

Where the Paris Settlement was perceived by some to be a foreign policy success and by others to be a final admission that the war amounted to a strategic blunder, there was near-unanimous agreement that Nixon's 1972 trip to China was a strategic success. In addition to formalizing relations with China, among the trip's outcomes was the opening of another axis of engagement for pressuring the Soviet Union.

The resulting era of triangular diplomacy yielded important opportunities to further American interests in Asia, but the most important policy initiatives would address European security. Partly this reflected increased concern that communist parties were gaining a foothold in Western Europe. NATO members Spain, Portugal, and Italy had large and growing communist parties that were perceived as a threat to Alliance stability. If communist parties grew significantly in strength and began to influence their respective national foreign policy agendas, NATO's resolve against Soviet adventurism might falter. Henry Kissinger dubbed 1973 "the Year of Europe" to signal that the Nixon administration was refocusing on the primary axis of East–West tension and seeking opportunities to bolster détente.

While the Nixon administration focused on the Year of Europe, defense analysts re-examined global security challenges to assess their effect on the East–West balance of power. By mid-decade, defense planning discourse would begin to shift away from a largely nuclear-centric discussion of strategic planning to one emphasizing conventional forces. Alain Enthoven captured the sentiment of defense planners in *Foreign Affairs*, writing that "if we really mean to maintain our nuclear guarantee as Europe's last line of defense, we must have strong conventional forces as a first line of defense."[5] Senator Sam Nunn echoed his views: "As long as the United States maintained a pronounced nuclear superiority over the Soviet Union at both the strategic and tactical levels, we could effectively deter conventional aggression. That superiority, however, has vanished, and with it the notion that NATO need not muster a credible conventional deterrent."[6]

Politically, conventional modernization would have to wait. "In the anti-military orgy spawned by Vietnam," Henry Kissinger recalls, "to have challenged the overwhelming Congressional sentiment for 'domestic priorities' was almost an exercise in futility, pouring salt on the open wounds of the Vietnam debate."[7] Funding for defense modernization remained a central issue. The Vietnam War had cost upwards of three-and-a-half billion dollars, sapping funds for procuring equipment, developing advanced weapons, or improving training. High rates of inflation throughout the 1970s made matters worse by lowering the actual buying power of defense dollars. Weapons systems were also more complex and more costly, leading to what defense economists termed cost growth. As weapons grew more complex, their per-unit cost increased.

John Lewis Gaddis summarizes Nixon's approach to defense planning in the immediate post-Vietnam environment as "consisting of (1) the appeasement of Congress, with a view to defusing as much as possible growing anti-military sentiment there, and (2) negotiations with the Russians aimed at restricting as much as possible their own military buildup, without constraining in any significant way comparable measures the United States might choose to take once the furor over Vietnam had died down."[8]

Secretary of Defense Melvin Laird spearheaded Congressional appeasement. A former Congressman reportedly skilled in "bureaucratic gamesmanship,"

Laird's chief mission was "forcing the American military to adapt to a harsher post-Vietnam environment without significant loss of either morale or capabilities."[9] Noteworthy here, given the press of conventional issues and declining defense spending, was that he sought "several new strategic weapons systems—the B-1 bomber, the Trident submarine, the cruise missile—but only at the price of a substantial reduction in conventional forces."[10] Not until the end of the decade would conventional issues receive Congressional support. For the time being, as Kissinger relates, the plan was, quite simply, to focus on preserving the "sinews" of American strength, in this case the pillars of nuclear deterrence.[11] Subsequently, nuclear weapons modernization was funded while overall military spending *decreased* "at an annual rate of 4.5 percent between 1970s and 1975"; this severely limited "the capacity of the United States to project conventional military power."[12] Even at mid-decade, "long after trends in Soviet military spending had become too obvious to ignore," the defense budget was cut an additional seven billion dollars as strategic nuclear programs received additional funding.[13]

In the mid-1970s, many feared that the American nuclear deterrent would be "decoupled" from Europe. This would leave Western Europe susceptible to Soviet nuclear blackmail or destine to be annihilated by theater nuclear exchanges that left the American and Soviet heartlands free of direct nuclear strikes. Such fears were aggravated by fears of new Soviet nuclear weapons.

Among the most important military developments of the post-Vietnam period involved missile systems, including the introduction of the SS-20 mobile intermediate range ballistic missile (IRBM) and the pursuit of a new class of advanced cruise missiles. The SS-20, deployed in 1977, was much superior to the missiles it replaced in virtually every measure of IRBM capability: reliability, survivability, range, accuracy, and the time required to fire, move positions, and reload. The SS-20 was the first Soviet IRBM armed with multiple warheads. Mobile missiles would lead to new US counterforce targeting requirements, which in turn led to new intelligence missions. Counterforce targeting required finding and destroying Soviet mobile missiles and related command and control capabilities before the missiles could be launched. Because the missiles were mobile, this necessitated the ability to dynamically change the target locations for bombers and missiles already in flight—capabilities discussed in Chapter 5.[14]

Soviet mobile missile advances spurred NATO to deploy US Tomahawk ground-launched cruise missiles (GLCMs) and Pershing 2 missiles, termed long-range tactical nuclear forces (LRTNF) because they had ranges between 3,000 and 5,500 miles. The initial LRTNF plan placed 96 cruise missiles in Germany, 160 in Britain, 112 in Italy, 48 in Netherlands, and another 48 in Belgium. West Germany's Helmut Schmidt was among those who viewed new US systems as crucial for NATO deterrence because their deployment signaled continued US commitment to a nuclear defense of NATO. West Germany would also host 108 Pershings, the highly accurate, high-speed missiles with a thousand-mile range that could reach targets in seven or eight minutes from launch.

A new nuclear arms race in Europe created domestic political problems for key NATO members. Mass political demonstrations erupted across Western Europe against the LRTNF deployments. East–West relations declined throughout the decade, muddying the arms control negotiations that embodied the last gasps of détente. Political pressure to find alternatives to nuclear deterrence intensified.

While missile developments altered the fabric of security affairs, the stark reality of Soviet conventional capabilities impelled a renaissance in American military thought and an overhaul of defense planning. Before the SS-20s were fielded, the October 1973 Arab–Israeli War had already begun to shape thinking about conventional deterrence alternatives. The October War fueled the quest for technological solutions to NATO's European security challenges and informed training and doctrine innovations. For the US Army, according to historian Richard Swain, the war "came as a shock to the Army because it pitted three mechanized armies looking much like those facing each other in Europe in a series of battles that suggested a revolution in military affairs had occurred while the Army was preoccupied with Vietnam."[15]

The October War demonstrated the effectiveness of precision-guided munitions (PGMs). Until the early 1970s, the high-velocity main gun on a tank was the only viable battlefield anti-tank weapon, and battle tanks were the coin of the realm in conventional force planning. Alternatives for defeating tanks were awkward to employ, inaccurate, packed too little punch, had too short a range, or, as in the case of jeep-mounted recoilless rifles, had noticeable back-blasts exposing the crew to enemy fire. Leaving aside the issue of weather and terrain differences between the Sinai Desert and Central Europe, the October War also suggested to some that inexpensive, man-portable, accurate tank-killers seemingly rendered US armor vulnerable to Soviet infantry forces. If tanks were no longer required to kill tanks, then American armor could be engaged by other Soviet weapons systems while the tanks pushed forward.

Planners comparing combat losses during the October War to projected losses on the plains of Central Europe were stunned by the high rate of munitions expenditures and rapidity of battlefield losses, calling into question the adequacy of NATO's pre-positioned munitions stocks and equipment levels. Combat attrition was indeed staggering. The Arab and Israeli armies lost more armor and artillery than the US Army had deployed in Europe at that time. The belligerents went through a staggering amount of ammunition. Some believed that, if similar munitions expenditures and attrition rates occurred in Europe, NATO would have to use nuclear weapons even sooner than anticipated.

In addition, the effectiveness of Soviet air defense systems caused NATO planners to rethink the role NATO's air-delivered tactical nuclear weapons would play in defending against a Soviet surprise attack. At the time, the US Air Force was assigned a lead role in blunting Soviet armored attacks. This required the ability to operate in the airspace directly over advancing Soviet forces long enough to deliver tactical nuclear weapons. During the October War, Soviet integrated air defense systems were even more capable against Israeli aircraft than those the North Vietnamese employed with alarming success against US aircraft. Where increased effectiveness of Soviet ground forces suggested that a surprise armored attack might cripple NATO's ability to use ground-launched tactical nuclear weapons, air defenses suggested that air-delivered tactical nuclear weapons might not be effective either.

In the aftermath of the October War, it was increasingly clear to a growing number of analysts that deterrence stability now depended on "a balanced military posture in which the deterrent value of each component—conventional, theater nuclear, and strategic nuclear—is magnified by its relation to the other two."[16] As the effectiveness of Soviet PGMs fueled debate over their effect on the correlation of forces in Central Europe, American defense planners called for PGMs to be integrated into NATO's arsenal. American forces armed with similar weapons systems, enabled by surveillance and targeting subsystems more suitable to weather and visibility conditions in Europe (not an issue in the Sinai), might blunt Soviet armor more effectively.

By 1975 weapons experts appeared frequently before Congress to testify about new precision munitions, intelligence capabilities, and weapons systems. Suggesting that new technology enabled "substitution of small weapons for larger ones," Henry S. Rowen surmised that "for many missions it may be possible for nonnuclear warheads to be substituted for nuclear ones" with the net effect of enhancing deterrence.[17] This appealed to those seeking to strengthen the defensive and retaliatory capabilities of US forces without increasing manpower or battle tank deployments. Some argued that precision munitions might provide an alternative to early use of tactical nuclear weapons, a possibility that appealed to European politicians as massive anti-nuclear demonstrations swept across Europe.

Conventional force modernization activities subsequently received newfound support. Shortly after Carter assumed office, NATO members agreed to a new long-term defense program (LTDP). Formally adopted in May 1978, the LTDP evolved into a fifteen-year modernization plan addressing ten mission areas. Only one concerned nuclear forces (Pershing II and cruise missiles). Others included air defense; anti-submarine warfare; anti-armored capabilities; advanced air-delivered munitions; command, control, and communications advances; improved reserve readiness and mobilization; and electronic warfare. Allies also agreed to increase annual defense spending and to jointly develop airborne early-warning aircraft. Pursuit of advanced conventional capabilities was reinforced in the 1978 NATO Summit communiqué, which highlighted the risk of a Soviet attack with minimal warning and related need for rapid reinforcement after hostilities commenced.

While defense analysts pondered the implications of new conventional weapons, the national security planning community was struggling to understand the implications of emerging economic issues. Concerns about economic stability increased with the 1974 announcement of a New International Economic Order of nations seeking to reduce Western dominance of global economic affairs. Western states became the target of a non-aligned political movement seeking to increase the collective political power of poorer, non-European states.

Economic security became much more important after the Organization of Petroleum Exporting Countries (OPEC) oil embargo rattled the American public and placed energy security at the heart of national security planning. The oil crisis also spurred important new regional security agreements, leading to plans for American military involvement in the Persian Gulf. Of note was a study conducted by Paul Wolfowitz during the Carter administration, the first serious consideration of Iraq's potential threat to American interests in the region. Three decades later Wolfowitz occupied the number two position at the Defense Department and was among the most vocal supporters of a military attack against Iraq to unseat Saddam Hussein from power.

American national security planners also grappled with international terrorism in the 1970s. US efforts to coordinate national counterterrorism policy began in 1972 with the creation of a Cabinet Committee to Combat Terrorism. In 1977, Jimmy Carter directed his Secretary of Defense Harold Brown to create Delta Force, the secretive Special Forces unit that assumed a lead role in the war on terrorism in the 2000s.

For these and other reasons discussed later, the 1970s were a time of intellectual innovation among military thinkers and technologists that, for decades, had been on the margins of mainstream defense strategy. Political developments in Europe, leadership changes inside the Kremlin, Soviet activities in the third world, and a growing Soviet global naval presence fostered "a political climate conducive to efforts to raise the theater-nuclear threshold through the improvement of conventional forces."[18] Raising the threshold remained a prominent theme in US national security planning throughout the late 1970s and early 1980s. One of the most important official studies at the time was a 1976 Defense Science Board study of solutions to offset Soviet military advantages. Drawing on an impressive array of experts and benefiting from other studies, it concluded that new technologies offered the potential to counter successive waves or echelons of Soviet forces in the event of a ground war in Europe.[19]

Toward technological innovation: DARPA and the offset strategy

History of technology scholar Alex Roland notes that the "Army concluded from Vietnam that it needed not less technology but more. It was not that smart weapons were bad; rather, they were not smart enough."[20] General William C. Westmoreland was among those that had already begun to think in earnest about how emerging technology would shape future battles. In early 1969, as American support for Vietnam plummeted, he outlined a vision in which "enemy forces will be located, tracked and targeted almost instantaneously through the use of data links, computer assisted intelligence evaluation, and automated fire control."[21] Using language nearly indistinguishable from persistent surveillance arguments made in the early 2000s, Westmoreland envisioned having "24-hour real or near real time surveillance of all types" and a force "built into and around an integrated areas control systems that exploits the advanced technology of communications, sensors, fire direction, and the required automatic data processing."[22]

A year later Congressional hearings on the "electronic battlefield" concluded that "the electronic or automated battlefield represented a whole series of technologies and programs that were combining to form a totally new American way of war."[23] Senator Barry Goldwater opined that battlefield information systems represented "the greatest step forward in warfare since gunpowder."[24]

The Defense Advanced Research Project Agency (DARPA) played an important role in bringing the power of emerging technology to bear on the Soviet conventional threat to NATO. Founded as the Advanced Research Projects Agency (ARPA) in 1958 and renamed in 1972, DARPA's work influenced the course of force modernization and military thought in areas as diverse as precision strike, sensor development, battlefield visualization, and automated targeting.

When it turned its attention to conventional warfare, DARPA had already proven itself extremely effective in solving critical national security challenges. The agency was founded in the wake of Soviet's 1957 Sputnik satellite launch to ensure "the United States would never be left behind in the area of new technology."[25] Throughout the 1960s, efforts addressed such priority issues as the space race, defenses against ballistic missiles, and the detection of foreign nuclear weapons tests.[26] From 1960 through 1965 missile defense and test detection work continued (accounting for some 70 percent of the budget) alongside new focus areas supporting counterinsurgency warfare (reflecting the importance of the conflict in Vietnam) and on computer processing. DARPA engendered "a fundamental revolution in integrated circuit design" that "had a major impact on computer technology."[27] By the mid-1980s, a combination of programs and initiatives— increasingly related in concept, doctrine, or operations—cohered, leading to the "implementation of disruptive capabilities."[28]

Director of Defense for Research and Engineering (DDR&E) Malcolm S. Currie appointed George Heilmeier to head DARPA in 1974, charging him to energize DARPA to "harness emerging technology capabilities to address the challenge of Soviet military" advances and evolve leap-ahead technologies to offset Soviet superiority in Europe.[29] Reinvigorated by new leadership and increased funding, the agency turned to " 'high risk, high potential payoff' " work in the mid- and late 1970s.[30] DARPA sponsored a workshop in 1974 to develop "a renaissance in conventional weapons technology and research" that helped foster an environment for other research and development efforts on conventional weapons.[31]

An emerging vision for the future of American military forces guided DAPRA's efforts. Richard Swain notes that, as the Army sought to revise its doctrine, training, and force structure after Vietnam, the vision Westmorland articulated in 1969 became the "new mantra." General DePuy revised it in 1974, predicting that in the future, "What can be seen, can be hit. What can be hit can be killed."[32] For current students of defense planning and military thought there is nothing profound in this. US planners seeking conventional solutions to perceived strategic and operational challenges in Europe during the 1970s believed it was a watershed in military history.

The research and development community benefited from detailed studies of Soviet military capabilities. Deputy Director of DARPA's Tactical Technology Office, Robert Moore, recalled that the office received "increasing amounts of information on the Soviet tank threats in Europe," including "regular intelligence briefings."[33] Among the important intellectual efforts was Joe Braddock's classified analysis of Soviet strategy, doctrine, and force structure. He identified potential weaknesses and suggested how to exploit them.[34] In 1976, a widely circulated Defense Science Board (DSB) Summer Study on conventional warfighting capabilities synthesized findings from other studies and provided an optimistic assessment of potential offsets to Soviet numerical advantages. The DSB endorsed the findings of an IDA Weapons System Evaluation Group's "target engagement study" and Air Force–DARPA work on the feasibility of "real-time targeting and missile guidance updates."[35] The Defense Science Board also endorsed Lincoln Laboratory's Integrated Target Acquisition and Strike System (ITASS) concept.

Testifying on the proposed budget in 1976, Currie told Congress that some 40 percent of the planned Fiscal Year 1977 research and development funds— more than four billion dollars—would be devoted to tactical issues. Investments, he argued, "reflected a transformation occurring in military technology" that would "change concepts and capabilities in command and control, mobility, armor/anti-armor, night fighting, massed firepower and the precision application of force at a distance."[36] DARPA's subsequent work had a significant impact on military effectiveness by fundamentally changing American capabilities to interdict enemy forces.

Richard Van Atta and Michael Lippitz argue that a broadly defined approach for leveraging emerging technology emerged by the end of the 1970s. Directed at the R&D community, the approach was a defense strategy to "offset" Soviet numerical superiority in Europe. Jointly devised by Carter's Secretary of Defense Harold Brown and William Perry, who replaced Currie as the DDR&E, the off-set strategy consolidated an existing, theretofore diffuse base of support for technology and organizational innovations.

Among the most important areas receiving attention was the operational requirement to interdict and disrupt second echelon forces, the waves of Soviet armored forces that would overwhelm NATO forces. Leland Strom, an expert on radar in Moore's office, proposed in 1975 "the concept of using an MTI (Moving Target Indicator) radar to track a missile to a ground target (e.g., group of tanks), 'close the loop' to guide it to the target and use terminally guided submunitions for the endgame."[37]

Perry recognized that activities begun by Currie and DARPA director George Heilmeier could increase the operational and strategic effectiveness of US conventional forces. In his words, doing so provided "qualitative advantages to American forces to offset the quantitative advantage the Soviet forces enjoyed" and later "achieved the status of a 'revolution' in military affairs."[38] In his 1978 testimony to the Senate Committee on Armed Services Perry outlined why precision strike advances offered the "greatest single potential for force multiplication" to meet the Soviet threat in Europe:

> Precision-guided weapons, I believe, have the potential for revolutionizing warfare. More importantly, if we effectively exploit the lead we have in this field, we can greatly enhance our ability to deter war without having to compete tank for tank, missile for missile, with the Soviet Union. . . . In sum, the objective of our precision guided weapon systems is to give us the following capabilities: to be able to see all high value targets on the battlefield at any time; to be able to make a direct hit on any target we can see; and to be able to destroy any target we can hit.[39]

Perry's testimony illuminates the essential vision of the "offset strategy," a term that in hindsight appears somewhat pedestrian in its straightforwardness because it had none of the unique symbolism or notoriety of other Cold War terms. Figure 4.1 lists several mid- to late 1970s technology initiatives that supported a vision Perry would champion for the rest of his government career.[40]

Substantively, the offset strategy spawned a technology investment portfolio yielding unprecedented returns in both military and nonmilitary applications. The research and development response to the Soviet threat was organized around the idea that several technology projects could be simultaneously matured and integrated into a larger system, a task that further brought systems engineering, systems integration, and information technology into the emerging vision for defeating successive waves of Soviet armored echelons.

See all high value targets on the battlefield at any time:
Airborne Warning and Control System (AWACS)
PAVE MOVER radar
Stand-Off Target Acquisition System (SOTAS)
Remotely Piloted Vehicle (RPV) and mini-RPV
Unattended Ground Sensors
Make a direct hit on any target we can see:
Army nonnuclear Lance missile and guidance advances
Army Patriot Missile
General Support Rocket System
Smart bombs
Destroy any target we can hit:
Rockeye bomb and bomblets
Wide Area Anti-Armor Munitions (WAAM)
Terminally Guided Submunitions

Figure 4.1 Precision strike projects.

Drawing on projects listed above, Robert Fossum—George Heilmeier's replacement as DARPA director—formally created the Assault Breaker technology demonstration program in May 1978. He recognized the potential to provide "a potentially step-level improvement in capabilities to redress the Soviet conventional threat."[41] The program entailed a high degree of risk due to its complexity, which included a number of systems integration challenges to achieve a networked architecture of sensors and shooters. Several subprojects in the demonstration were themselves very challenging, making Assault Breaker an extraordinary effort.[42]

"The question to be answered," Richard Van Atta relates, "was whether developments in sensors, computing, communications, guidance, and munitions allowed for deep precision attack against hard, mobile targets."[43] Among the other elements of the larger system for testing the concept was an attack coordination center able to fuse data from multiple sensors. Testing of key components began in the fiscal years 1979 and 1980 with critical tests occurring in the early 1980s. In late 1982, five terminally guided submunitions "made direct hits, one on each tank in a pattern of five stationary tanks."[44]

Assault Breaker reflected a fundamental shift in military strategy following Vietnam, a growing "realization that the timely use of tactical nuclear weapons to stop an attack in [Europe] was unrealistic."[45] The program leveraged developments in sensors, new weapons platforms, and advanced munitions "to greatly mitigate or potentially negate the Soviet threat and do so without resort to the use of nuclear weapons."[46] Assault Breaker aimed to test the capabilities outlined in Perry's testimony: to see all high value targets on the battlefield at any time; to be able to make a direct hit on any targets we can see; and to be able to destroy any target we can hit. Innovations later cohered into a new core competency organized around the idea of long-range, stand-off, precision strike with terminally guided submunitions. Soviet observers viewed the resulting capabilities to be a conventional variant of what they dubbed a theater strategic offensive, with the primary increases in effectiveness coming from a reconnaissance–strike complex.

Meanwhile, other DARPA projects demonstrated the power of emerging information technology to improve battlefield intelligence and targeting. Key projects discussed in later sections included the Coherent Emitter Location Testbed (CELT) and the Battlefield Exploitation and Target Acquisition (BETA) initiative.

Assault Breaker remains an important case for students of military innovation. It was among the Cold War's most ambitious systems integration efforts, although some of the program's ambitious goals remained unrealized until the early 2000s. In the end, it demonstrated a "a capability to attack multiple tank targets using terminally guided submunitions released from a standoff 'missile bus' controlled by an airborne radar."[47]

Service cooperation was critical. As Van Atta, Nunn, and Cook conclude, perhaps "even more important than the testing and developing of specific technologies is the conceptual breakthrough in getting the Services to work together across the barriers of roles and missions to attack the Warsaw Pact tank threat."[48] Assault Breaker's management structure was based on a joint program office that struggled to address often diverging interests of Service sponsors. Army and Air Force collaboration was in part facilitated by the support of Generals Don Starry, commander of the Army's Training and Doctrine Command (TRADOC) and William Creech, head of the Air Force Tactical Air Command. Additional Army and Air Force support was due to the fact that Assault Breaker did not threaten the continuation of either services' existing programs.[49]

The first combat testing of the essential conceptual and technological elements of Assault Breaker occurred in the Persian Gulf, not on the plains of Central Europe. During the Gulf War, for example, some thirty-two Army Tactical Missile Systems (ATACMS) were used in conjunction with J-STARS. As Paul Nitze remarked about advanced conventional weapons developed for the defense of Europe, the "Gulf War offered a spectacular demonstration of the potential effectiveness of smart weapons used in a strategic role."[50] Technology alone does not render new, smarter weapons so effective that their mere existence evokes discussions of an extant revolution in military affairs. How they are employed, the organizational and operational innovations that enable their effectiveness as part of a warfare being waged by distributed units acting in concert, is the more important dimension of "change" we must consider.

Although not formalized as a defense department program in the same guise as flexible response or massive retaliation, Assault Breaker and other programs pursued under the Offset Strategy, including Stealth aircraft and communications advances, nonetheless shaped modernization decisions in the late 1970s and became the de facto principle underlying the Reagan defense build-up in the early 1980s. The Offset Strategy was an early expression of what current defense planners call "system of systems" thinking. Through a "synergistic application of improved technologies" that spanned "electronic countermeasures, command and control (communications, data links, and networks), stealth, embedded computers (microprocessors), and precision guidance (advanced sensors)," the Offset Strategy would "allow the U.S. to overcome Soviet defenses and destroy Soviet tank legions."[51]

Capabilities conceived during the 1970s provided the pivot around which flexible response could be reconstituted. This is a key chapter in the thirty-year transformation that began in the shadow of Vietnam. Admiral Bill Owens (retired), former Vice Chairman of the Joint Chiefs of Staff, argues that, if one were to "affix a date" to the origins of revolutionary advances in American military effectiveness "it is 1977, when three key Pentagon officials—Harold Brown, Andrew Marshall, and William Perry—began to think in concert about the application of technology to military affairs."[52] They did so, arguably, because they were acutely aware of the Soviet threat in Europe and well informed of how ongoing research and development activities could meet it. Existing systems would be improved through the use of better information technology. Information technology was subsequently leveraged to integrate new surveillance and reconnaissance capabilities with precision guided weapons.

An emerging revolution in army doctrine, training, and operational art

Reflecting back on the Army's turn from the rice paddies of Southeast Asia to the plains of Europe, General Donn Starry recalls that "the Soviets had been very busy while [the Army] was preoccupied with Vietnam" and had "embraced the notion that they could fight and win at the operational level of war with or without nuclear weapons. Their preferred solution: without."[53] Even if Soviet forces had in fact adapted their thinking on nuclear weapons, NATO's doctrine of flexible response and US declaratory policy concerning nuclear weapons use assumed that NATO *would* employ tactical nuclear weapons to stop a Soviet conventional assault.

Based on command assignments in Europe, Starry did not believe the decision to use tactical nukes could be made in time to prevent Soviet armored forces from racing across the west German plains into NATO territory. Similar sentiments abounded, giving rise to debate about Soviet capabilities and intentions. Bipartisan concern swirled over the reliability of US and NATO warning systems to uncover signs of an impending attack.

The Army developed new doctrine, training regimes, and technology that supported the new operational imperative to seize and maintain the initiative in the highly dynamic and increasingly intense battlefield environment suggested by the October War. The Army's adaptation to new strategic and operational realities initially focused on developing the ability to fight outnumbered and win the opening battles of a war in Europe without resorting to nuclear weapons use. By the mid-1980s, the same offensive approach would guide Army planning for non-European contingencies. In the 1990–1991 Gulf War, Army commanders who had matured in the post-Vietnam era indeed demonstrated an aggressive, flexible approach to battlefield leadership. Brigade and below (colonel and below) commanders displayed the greatest ability to adapt; they had benefited from new training regimes and matured within a cultural environment that was highly supportive of innovation. Some division-level commanders and most Corps and above commanders were less comfortable with fluid decision making and the greater independence demonstrated by more junior offices. In part, this was because training, command and control, and decision support systems had not developed the sophistication required for larger, Corps-level operations.

Surveillance, targeting, decision support, and weapons guidance systems matured in concert, leading to greater levels of command and control sophistication and improved coordination at greater battlefield depths and breadths, a trend that persists today as operations are increasingly global in nature. The drive toward precision, accuracy, and increased timeliness emerging from the 1970s operational challenges would evolve into the "rapid dominance" and "rapid decisive operations" schools of thought in the early 2000s. During the 1990s, the Army refocused the training provided to senior commanders to develop the same level of comfort with increased flexibility and agility that battalion and brigade commanders demonstrated during the first Gulf War. Decision making at division level and above was much more flexible during the 2003 invasion of Iraq. Throughout, successive doctrinal innovations and exercises facilitated the integration of new technology and operational concepts.

Among the most important developments in the Army's post-Vietnam renewal was a commitment to improved standoff precision strike capabilities. Army units had to be able to quickly identify, track, target, and destroy both stationary and moving enemy forces before they could directly engage American forces. Among the Army's R&D efforts along these lines was the helicopter-mounted Stand Off Target Acquisition System (SOTAS), an airborne targeting system similar in concept and operations to the Air Force's E-3. Precision strike required precision munitions, leading to the Air Force's wide area anti-armor (WAAM) project and the Army's Terminally Guided Sub-Munition (TGSM) designed for rocket systems and artillery. The Lance nuclear missile was adapted to the General Support Rocket System and eventually the Army Tactical Missile System; a Multiple Rocket Launcher System (MRLS) was also used in the first Gulf War. By 2003, the Army's digitized indirect fire support system was so efficient that it could clear a fire mission within minutes, ensuring that no friendly air or ground units were in the line of fire.

In a recent history of the US Army's armor branch, former General Don Starry summarizes the planning environment in the early 1970s, the dawn of the Army's digital age. "It was an era characterized by the expanded threat in Europe, a growing threat of conflict in the Third World (especially the Middle East), increasing worldwide economic interdependence, greater difficulty articulating political goals for the planners who designed military activities to achieve them, and intrusive and abrasive media probing into all aspects of military operations."[54] As America withdrew from Southeast Asia, "funds available for Army general-purpose procurement" in 1973 "were about two-thirds of the last pre-Vietnam year in terms of equivalent purchasing power."[55] This worsened long-standing readiness problems.

In this environment, it was impractical to pursue an expensive transformation program, no matter how desperately the Army needed a boost to recover from the Vietnam era. Vietnam, furthermore, had exacted the greatest toll on the Army in terms of delaying modernization efforts. According to one assessment, even during Vietnam the Air Force and Navy "developed, or were in the process of developing, new aircraft, new air-to-air missile weaponry in the air defense arena, and with dramatic new technologies were beginning to extend their reach into space."[56] Preoccupied in Southeast Asia, the Army had failed to adopt new technology, adapt doctrine, and modernize training.

At mid-decade strategic and operational challenges in Europe increased the necessity for reform and transformation. Many of these challenges were similar in concept to those prompting flexible response debates in the 1960s. The principle difference was the fact that deterrence credibility problems now occurred in a strategic environment that included parity in nuclear forces and a sizeable Soviet advantage in conventional forces. By the late 1970s, flexible response no longer had the credibility required to serve as a NATO strategy of deterrence. For ground force commanders, theater planning and readiness challenges became more diffuse. For defense strategists, it appeared that strategic planning for European contingencies was increasingly coupled to peripheral regions. Military strategists renewed their support for doctrinal and technological innovations that could reduce timelines for warning that an attack was imminent, for quickly transitioning to the defense, and for enabling a rapid shift to a more offensive or retaliatory posture.

As General Haig observed, however, the relative lack of attention given to forces in Europe during the Vietnam war "tended to breed a garrison mentality" that had a negative influence on readiness, one that was "especially marked among our ground echelons" which, "unlike their air and naval brethren" did not "routinely function in an environment of high operational intensity."[57] Rick Atkinson's definitive account of the post-Vietnam Army, *The Long Gray Line*, paints a more depressing picture. US Army forces in Europe were "bored and ignored" by defense leaders during Vietnam at the same time that it were "bled white to keep the US war machine in Southeast Asia supplied with officers, experienced NCOs [non-commissioned officers], material, and money."[58]

A related challenge involved the quality and morale of Army personnel in Europe. Weekly press reports of murders, rapes, riots, robberies, drug-related incidents, and other crimes in and around US military bases reflected poor morale as much as poor leadership. In one incident, German firefighters refused to respond to a fire at an American military base because they feared being attacked and beaten by US soldiers. Drug use was reportedly just as high among regular infantry troops as it was among the technicians responsible for nuclear weapons.

Personnel challenges were not confined to Europe. An additional concern was the status of soldiers inducted under Lyndon Johnson's Project 100,000, which brought some 300,000 underprivileged and unemployed recruits into the military as part of the Great Society program. The program mandated lower minimum standards for intelligence and physical health; many of the recruits were illiterate. Post-Vietnam studies revealed that Project 100,000 inductees were killed in action at nearly twice the rate of other combat soldiers. By 1973, many had progressed into the non-commissioned officer ranks, entrusted with leading and training the backbone of the Army: its soldiers. Recruiting ills were aggravated by a defense department decision mandating that a quarter of the Army's recruits be high-school dropouts, a decision intended to bolster recruiting numbers.

Army Chief of Staff General Creighton W. Abrams revamped the way the Army prepared for battle. He did so despite dwindling political support for the funding of new weapons systems and a growing political aversion to any increase in readiness that appeared to be preparing the Army for another regional conflict. In July 1973, he also created a Training and Doctrine Command (TRADOC), tasking the new organization to develop doctrine and to define the Army's emerging combat mission in a way that would gain support for a more sustainable modernization program. Headquartered at Fort Monroe, Virginia, TRADOC quickly assumed responsibility for the rebuilding of the Army, developing a new approach to warfare, shaping force structure, and for writing a comprehensive set of doctrinal guides.

Lieutenant General William E. DePuy was tapped to be the first TRADOC commander. DePuy's role in setting the stage for an overhaul of Army training and for establishing new possibilities for doctrinal innovation can be likened to wiping the slate clean for new thinking. If significant or major innovations are associated with the creation of innovation streams, meaning the institutionalization of innovation and disruptive thinking, then the work of TRADOC leadership in the 1970s surely amounts to one within the Army. Earlier, the Army decided to pursue five weapons systems considered key to its long-term force modernization objectives. The so-called "big five" included a new main battle tank, an armored infantry fighting vehicle able to support the tank on the battlefield, an attack helicopter, an assault helicopter able to carry troops, and an air defense system. DePuy was a key figure in the big five decisions and considered one of his tasks at TRADOC to be preparing the Army to organize and train with the new weapons.

The consequences of failing to adapt during the Vietnam years had been highlighted by the effectiveness of new weapons during the 1973 October War. New precision guided munitions and air defense systems making their combat debut on the Golan Heights and in the Sinai Desert challenged prevailing assumptions about armored warfare. The Israelis were able to fight outnumbered and win, some concluded, because they possessed superior tactical doctrine and better training. The organization directed to revamp post-Vietnam training had barely unfurled its command flags when the October War shocked Army leadership with the implications of anti-tank and air defense technologies for war in Europe. As TRADOC was created specifically to help rebuild the post-Vietnam Army, understanding the October War became a priority for Abrams, who ordered TRADOC to undertake intensive studies of all aspects of the conflict. He wanted to understand what new technology and operational concepts implied for a potential fight against the quantitatively and qualitatively superior Soviet Red Army.

TRADOC's Special Readiness Study Group, headed by Major General Morris J. Brady, visited the battlefields of the October War and gleaned some 162 Army-specific issues for consideration. Three overarching conclusions emerged. "First, the battlefield environment was far more lethal than ever before. Second, fighting demanded a highly trained and integrated combat arms team. Third, tactical training could make the difference between success and failure."[59]

The October War also "added new momentum" to existing Army–Air Force initiatives "to adopt a more joint approach to airland combat" and pursue "a set of complementary capabilities that any potential enemy would find difficult to match."[60] Accordingly, a Joint Army–Air Force Studies Group was formed at Nellis Air Force Base in June 1975; a month later the Air–Land Forces Application Directorate (ALFA) was created at Langley Air Force Base in Virginia. ALFA would oversee a critical Joint Second Echelon Interdiction Study in 1979. Based on these and other air–ground concept development activities and studies, a 1983 agreement for the Enhancement of Joint Employment of the AirLand Battle Doctrine was signed between the Air Force and Army.

Brigadier General Paul Gorman was a TRADOC founding father and among the most widely recognized members of the generation of notable leaders that rebuilt the Army. His unique contribution was radically changing Army training. Gorman personally supervised the revision of Army training literature and, as TRADOC's chief of staff, intervened to ensure that standards and skill qualifications were overhauled. Decrying the lack of realistic training environments, especially in Europe where political concerns severely limited live-fire exercises and large-scale maneuvers, Gorman fought for what became the National Training Center (NTC) in the desert at Fort Irwin, California.

NTC was large enough to conduct exercises using the latest weapons at realistic ranges and to incorporate air–ground training. Instrumentation to document training in high-tempo scenarios improved the learning process. Army units trained in realistic maneuvers against NTC's opposing force or "OPFOR," which was chartered to identify and exploit any weaknesses in visiting forces, the Blue Force. The OPFOR wore Soviet-style uniforms, carried Soviet weapons, and were experts in Soviet operational tactics. Some maintained that NTC's OPFOR was better at Soviet tactics than any Red Army motorized rifle regiment. Visiting units were often so humbled by OPFOR victories that fights would break out when the OPFOR would begin to provide an after-action critique of Blue Force mistakes. Blue Force commanders and senior NCOs were frequently dismayed at the critiques levied by more junior OPFOR members. The Army was still unaccustomed to open criticism. It would take decades for a "learning organization" mentality to develop and for senior leaders to actively encourage their subordinates to voice ideas or alternative approaches to tactical challenges.

After putting in motion a radical change in how the Army trained, General Abrams pursued a similarly radical shift in how active, reserve, and National Guard forces would prepare for and be called into service. Abrams obtained a commitment to increase the number of Army divisions from thirteen to sixteen in a 1973 deal with then Defense Secretary James Schlesinger that Pentagon insiders dubbed the "Golden Handshake." In return for the increase, Abrams agreed that the Army would staff the new divisions from existing manpower. In other words, the increase in divisions would not require additional personnel. Each division, the Army's basic organization for wartime mobilization, contained some 16,000 people. To generate the manpower needed for the new divisions, Abrams unilaterally decided to transfer combat support units from the active Army to the Army Reserves and National Guard. Some 70 percent of the Army's overall support capabilities would eventually move from the active Army to the Guard and Reserve.

This "Total Force" concept meant that the Army would not be able to go to war without National Guard and Reserve units being mobilized. Because political leaders would have to commit reserve units early in a conflict, Abrams believed that politicians would not send the Army into battle without having to build support for the war itself and, in doing so, sustain support for the forces involved. Army reserve requirements remain particularly high, witnessed by the number of soldiers called to active duty in the 2000s to simultaneously fight a global war against terrorism and remove Saddam Hussein from power in Iraq.

Along with new training capabilities and additional active duty troops, the Army developed a new capstone warfighting doctrine. TRADOC spearheaded the creation and evolution of a new field manual, FM 100-5, *Operations*. First published in 1976, FM 100-5 underwent revision in 1982 and 1986. Revisions are discussed in Chapter 5. The 1976 manual espoused an Active Defense doctrine and reinforced visions for an Army that placed initiative at the core of its organizational culture. Thoughts on Active Defense were shaped by detailed studies of nearly a thousand battles involving numerically uneven forces. Reflecting the Army's concern with Soviet numerical superiority, the studies examined cases where forces fought outnumbered and prevailed in battles, in localized areas of a larger front, or in skirmishes. Analysts concluded that, with the right training and technology, smaller forces fighting outnumbered with as much as a six to one disadvantage could indeed seize and maintain the initiative throughout a battle.[61]

While commanding the Army's V Corps, Starry encountered an institutional malaise that seemed to discourage commanders from pursuing innovations in training. Commanders were not taking the initiative in their personal lives, in local training, or in exercises to test their readiness for combat. Few subordinate maneuver commanders were familiar with the terrain they were to defend, a deficiency he sought to overcome by leading intensive staff rides across NATO's Central European front. How could commanders seize and gain the initiative if they didn't know the terrain? They had no organizing construct or doctrinal framework to approach their preparations for battle along the Central European front. How could new leaders be educated to successfully operate at the operational level of warfare, where successfully leading corps-level units required a deeper understanding of the principles of warfare and the art of command?

Active Defense rested on the imperative to see deep into the enemy's rear to locate successive waves or echelons as they moved forward toward NATO's front lines. To offset the numerical advantage of forces arriving at a broad front, the Army had to maneuver quickly to concentrate its forces for a defense while simultaneously striking advancing forces before they could punch through NATO's defenses.[62] Strict adherence to the principle of "economy of force" was required along with the ability to absorb the initial attack, channel it, and then launch superior counterattacks at key moments against weaknesses. For the countless US military theorists rediscovering Clausewitz in the early 1980s, this reflected the Prussian's concept of a "culminating point."

The 1976 manual was criticized almost immediately upon publication. Congressional staffer William Lind published a damning, systematic critique of Active Defense in *Armed Forces Journal*. Among the points Lind raised was the Doctrine's glaring lack of attention to winning the "second battle," or defeating the echelons of Soviet forces after stopping the initial attack.[63] DARPA programs were indeed addressing the need to simultaneously attack successive echelons of armored forces. Doctrinal changes to address the problem are discussed in Chapter 5.

Active Defense had been refined during war games sponsored by the Army's Command and General Staff School. The games were designed to explore alternatives for a defense of NATO. But upon publication, Active Defense doctrine was poorly understood even among those assigned to teach it. Unclear articulation of precepts left readers questioning how to implement the doctrine, leading to a subsequent 1978 manual on implementation and execution. Starry himself found it difficult to teach Active Defense to his staff when he served as Commander of V Corps in Europe from February 1976 through June 1977.

Others soon joined Lind, resulting in a wide and varied assault on Active Defense. One historian described it as "the most read and most attacked doctrinal statement in the history of written doctrine in the U.S. Army."[64] A constructive, albeit heated debate ensued. Main lines of dissent included the feasibility of advanced technology as a solution, the proper Army division and corps structure for achieving agility, the efficacy of different defensive postures (e.g. forward defense versus a defense in depth), and the optimal mix of armored and light infantry forces. Interlocutors all had the same objective: fixing shortcomings in American military doctrine.

By the end of the 1970s, much thinking had indeed occurred; the Army was already on its way to a post-Vietnam renewal. To their credit, senior leaders welcomed debate, perhaps recognizing FM 100-5's flaws or its status as an interim step; however hesitantly, at the very least the manual focused dialogue on the future of the Army. A transitional document in the evolution of military thought during the late 1970s, FM 100-5's evolution sparked "some of the richest professional dialogue in the U.S. Army's history."[65] In 1979, even as Army Chief of Staff Edward C. "Shy" Meyer directed Starry to revise the manual, Meyer noted that the manual wrought "profound and widespread dialogue across the entire spectrum of basic tactical doctrine" and "caused people to think aloud for a change."[66]

Air power developments: precision warfare, tactical aviation, and space power

At one point during the Vietnam War, the Air Force reportedly ran out of bombs and was forced to buy some five thousand bombs back from the West Germans, who had bought them as scrap. Conventional munitions were simply not a priority for an institution that considered itself a strategic nuclear force. By the end of the 1970s, however, critical changes would occur in Air Force doctrine, operational tactics, and force planning that yielded advances in air–ground battlefield cooperation in the 1990s and 2000s.

The 1971 publication of a revised Air Force Manual (AFM) 1-1, *United States Air Force Basic Doctrine*, is a benchmark in the evolution of Air Force thinking about what would become known as AirLand Battle doctrine.[67] It posited that nonnuclear conflicts, including regional wars involving Soviet clients, required "sufficient general purpose forces capable of rapid deployment and sustained operations."[68] Going beyond earlier flexible response discussions, AFM 1-1 stated that *all* elements of the Air Force "are responsible for conducting and supporting special operations."[69] Of note is the lack of specific mention of support to all ground operations. Yet the doctrinal assertion that general purpose forces were needed for rapid deployments and that air power needed to support special forces would be taken to an extreme thirty years later when B-52 bombers designed for nuclear war directly supported special operations troops on horseback in Afghanistan against the Taliban.

But the 1970s did not witness the same level of doctrinal innovation and debate within the Air Force as that occurred in the Army. The 1979 AFM 1-1 retained the strategic nuclear mission as the Air Force's highest priority and reinforced a traditional antipathy for using strategic bombers for nonnuclear missions. Some later referred to the 1979 version as a comic book—it had many pictures, was written in a large font, and did little to advance air power doctrine.

Doctrine, perhaps, was not as important to the Air Force as it was to the Army. The Air Force traditionally focused more on technology and systems than theory and doctrine. Pilot training was more about tactics and mission profiles. At the wing level the training and focus tended to be on individual performance, not on integration. Army battalion and brigade commanders, meanwhile used doctrine to provide a common organizational framework to ensure that large, distributed forces had similar views on tactics and the employment of weapons against the enemy. Air Force leadership was less concerned with doctrine than developing the organizational competencies required for new precision strike missions, to use electronic countermeasures and electronic warfare capabilities, to revamp close air support to ground forces, and to address new strategic airlift requirements.

New roles and missions contributed to changes in Air Force thinking about its core competencies, leading to changes in the Air Force's institutional priorities in the 1980s and, eventually, to the development of the first comprehensive Air Force doctrine in the 1990s. Important here were how initial thinking about roles and missions changed resource allocation decisions in the 1970s, shifting the emphasis to nonnuclear missions. Among the most important but frequently overlooked missions were strategic airlift and in-flight refueling of fighter aircraft, tankers, and heavy lift aircraft. Tactical combat missions also received additional funding. "Within the air force," military aviation historian Richard Gross reports, the Strategic Air Command's (SAC) "share of the budget and force structure declined significantly while its tactical forces gained in relative importance."[70]

Personnel changes were an early indication of larger cultural changes. Fighter pilots—not bomber pilots—began rising to senior positions. By the early 1980s, "there were no longer any bomber generals in senior Air Staff positions."[71] Another indication of the shift concerned force structure. During the 1970s, the number of medium bombers assigned strategic missions increased from 4 to 60; heavy bombers decreased from 465 to 316. The number of medium bombers assigned to tactical support missions leaped from 26 to 264. If force posture reflects doctrine and planning, it appears that Air Force did indeed undertake a transition in its operational focus during the 1970s. By the mid-1980s, B-52 strategic bomber crews began training in earnest for conventional warfare.[72] By the late 1990s, conventional training was ubiquitous.

The definition of strategic airpower was also changing. According to Institute for Defense Analyses scholar Richard Van Atta, some Air Force leaders "quietly focused on an emerging 'grand challenge' in the 1970s, the ability to strike any target, any where in the world, day or night, with precision."[73] Early thinking about a global, expeditionary air force was motivated by Soviet expansionism outside of the European theater and the recognition that a narrow, nuclear-centric view of airpower would constrain the Air Force's ability to contribute to a global deterrence strategy.

An associated development involved technology to support global targeting and navigation. Precision location and navigation advances were essential to airpower's evolving NATO defense missions as well as the emerging vision for global conventional strike. Important here is the fact that conventional innovations were possible because of investments made in nuclear planning and delivery systems. Among them was the development of a global geodetic referencing system. Because the earth is not perfectly round, and because magnetic anomalies interfered with some navigation equipment, it was difficult to accurately deliver a missile from one hemisphere to another. Until the 1970s, strategic targeting required a massive effort to survey reference points around the globe and to develop math models to guide navigation. Advances in geolocation during the 1970s enabled the development of a coordinate system referenced to the planet's center, allowing for easier adjustments to correct local anomalies. New navigation technologies included a ring-laser gyroscope, better timekeeping devices, and other advances facilitated by the advent of microelectronics and miniaturization. Capabilities developed to support global nuclear strikes were adapted to conventional missions.

Global strike also required global command and control capabilities. In April 1973, the Air Force was designated the lead Defense Department agency responsible for integrating Air Force, Navy, and Army satellite navigation programs into a single development program, initially called the Defense Navigation Satellite System. By September, a Joint Program Office led by the Air Force and staffed by representatives from all the military services reached a compromise on a program that adapted satellite orbits from the Navy's Timation system and signal frequencies from an Air Force proposal. The resulting system evolved into the Navstar Global Positioning System (GPS), a critical program in the thirty-year transformation in American military effectiveness.

The Navstar program focused on the procurement of satellites and development of user interfaces from 1973 to 1979, including the testing of ground navigation equipment using simulated airborne signals in 1977. An Altas rocket carried the first GPS Block-I satellite into space a year later. Three subsequent 1978 launches delivered the world's first three-dimensional global positioning capability. Four additional satellites were launched from 1979 through 1985, completing phase two of the GPS program. Prototype GPS receivers began using live satellite data in 1982, leading to the extensive use of the GPS signal. Navstar GPS reached full operational capability in 1995 with twenty-four Block-II satellites in orbit. Noteworthy is the fact that advocacy for GPS came from the

Office of the Secretary of Defense and senior civilian leaders, not the military services.[74]

Air defense and air superiority missions were also revamped in the 1970s. During the 1973 October War, Egyptian and Syrian forces inflicted heavy losses on the Israeli air force, which was unprepared to counter the effectiveness of the Soviet Gainful SA-6 mobile surface to air missile (SAM) system. Countermeasures to SAM radar detection and tracking capabilities involved overpowering, spoofing, or jamming the radars themselves. Such electronic countermeasure (ECM) techniques had been central to air power developments since Second World War. But Soviet innovations were occurring faster than NATO countermeasures, leaving open the possibility that additional SAM developments would provide the Soviets with a window for technological surprise. Without air power, NATO forces would be even more vulnerable to a massive armored thrust.

Another way to defeat enemy radar and associated air defenses was to reduce the aircraft's radar cross section (RCS). In 1974, DARPA asked General Dynamics, Northrop, McDonnell Douglas, Grummen, and Boeing, the largest US military aircraft companies, to study the feasibility of an aircraft with an extremely small RCS and to build test airframes to demonstrate design options. Lockheed was not included, but later joined and eventually won the competition to design and prove the feasibility of a low-RCS aircraft able to penetrate Soviet airspace without being picked up by Soviet radars.

DARPA-sponsored studies actually led to two stealth programs that defense analysts in the early 2000s considered cornerstones of a new American way of war, the Lockheed F-117 and the Northrop B-2 stealth bomber. Motivations for both stemmed from US losses incurred during a mission to bomb Hanoi in which 5 percent of the B-52s were lost. Soviet air defenses were much better over the Soviet Union. They reportedly extended to some 125,000 feet with overlapping, integrated radar coverage. Mobile air defenses demonstrated in the October War further complicated the task of penetrating Soviet airspace, increasing fears that even tactical aircraft assigned to deliver conventional or tactical nuclear strikes against Soviet armored units might fail.

Lockheed's success in developing what was later termed stealth aircraft was in large part due to the work of mathematician Bill Schroeder and software engineer Denys Overholser. Schroeder based his work on nineteenth-century Scottish physicist James Clerk Maxwell's equations on the propagation of energy reflected off surfaces, German engineer Arnold Johannes Sommerfeld's radar reflectivity studies, and Soviet scientist Pyotr Ufimtsev's work in the early 1960s on electromagnetic reflections. While each of these works informed the effort, none of the equations were suitable for the more complicated task of designing an aircraft.

Schroeder's innovation was to reduce the problem to a series of two-dimensional surfaces—an aircraft design consisting of flat plates or facets that reflected the echo (or bounce) of the radar beam away from the radar. To pick up the target, energy from the radar had to bounce from the objective back to the radar system's receiver. Schroeder and Overholser collaborated on a computer program that provided RCS assessments of aircraft designs by sifting through different designs on which different plate or faceting approaches were applied. This was the first time that an aircraft would be developed by a computer program. Based on their work, Lockheed won the DARPA design competition.

The resulting aircraft program, named HAVE BLUE, "was a quarter-scale proof-of-concept aircraft designed to test out industry concepts of 'very-low-observable' capabilities while meeting a set of defined operational requirements."[75] After successful test flights in 1979 the program was accelerated as part of the offset strategy: "The DARPA stealth program was immediately transitioned to a Service acquisition program with an aggressive initial operating capability (IOC) of only four years—foregoing the normal development and prototyping stage."[76] Initial delivery of F-117s occurred in 1982 and fifty-nine were in service by the end of the Cold War.

Where Assault Breaker involved the integration of several technologies and weapons systems into a concept demonstration that informed subsequent systems and operational concepts, the development of Lockheed's F-117 Nighthawk stealth aircraft was a highly compartmented development project resulting in a relatively rapid transition from an operational prototype to production and fielding.

Air Force leadership supported the DARPA program responsible for the F-117 only after being assured that it would not compromise funding for the new F-16 strike fighter.[77] This was another case in which the Air Force did not actively pursue innovative new capabilities—the program was pushed by Dr Currie when he was the Director of Defense for Research and Engineering (DDR&E).

No survey of air power developments during the post-Vietnam period would be complete without mentioning attacks on two bridges, the Paul Doumer and the Thanh Hoa.[78] These were key bridges in the air campaign, but their destruction was really more important to the Nixon administration's larger strategy of sending the North Vietnamese a message during negotiations. By destroying them, the United States was communicating to Hanoi that American airpower could successfully strike important command, control, and logistics targets. Other strikes on military infrastructure demonstrated that American air strikes could cripple the North Vietnamese economy. While the bridges were not crucial to the overall war effort, they remain storied events in the history of precision bombing.

By 1972 the Air Force and Navy had flown over eight-hundred sorties and lost eleven aircraft trying to destroy just the Thanh Hoa. Why mention the bridges? They are part of the early history of American precision bombing. Among the previous attempts to destroy the bridges was a 1965 attack on the Thanh Hoa where over one hundred aircraft sorties delivered some five hundred bombs. Repeated attacks with "dumb bombs" and early generations of precision munitions failed to destroy the bridge. Things changed with the use of precision bombs and new attack profiles during the 1965–1968 Rolling Thunder campaign; decisive results occurred after munitions and tactics were perfected during the 1972 Linebacker I and II campaigns. During the Linebacker campaigns, both bridges were destroyed using two different types of highly accurate, innovative conversion kits that fitted guidance packages, fins, and other upgrades to regular bombs.

The first was the AGM-162 Walleye electro-optically guided high-explosive bomb, which had a small television camera mounted in its nose that transmitted an image to a weapons officer. After the weapons officer selected an aim point and locked it into a computer, the bomb guided itself up to the target (assuming the guidance system kept its lock) while the plane departed; it was a "fire and forget" weapon with an eight-mile standoff range. Walleyes, developed by the Navy, were the first precision weapons to prove their merit. In the final stages of Rolling Thunder, the Thanh Hoa was temporarily closed to traffic by Walleye hits. Out of sixty-eight Walleyes used against bridges, power stations, and military facilities during Rolling Thunder, all but three were direct hits. The Walleye's guidance system, however, required a high-contrast aiming point to be effective, meaning the bomb was less accurate in poor weather or if the adversary used smoke or camouflage.

A second, more revolutionary precision bomb was the Paveway laser-guided bomb, employed in routine strikes for the first time during the Linebacker campaigns. "Pave" stood for precision avionics vectoring equipment. Instead of a camera, its nose contained a sensor designed to lock onto a low-powered laser beam "illuminating" the target. In May 1972, twelve F-4 Phantoms, eight carrying two thousand-pound bombs each and the other four flying protective cover, struck the western span of the Thanh Hoa with a Paveway, pounding it off its supporting beams and rendering it unusable. The Paul Doumer bridge, the longest in Vietnam, was successfully attacked soon thereafter. No planes were lost in either attack.

Noteworthy in the Vietnam bridge attacks was the ratio of attack to supporting aircraft. As George and Meredith Friedman conclude, from the mid-1960s to the early 1970s, the "percentage of attack aircraft had shifted from 91 percent to 37 percent; in fact, the bulk of the aircraft that flew missions in 1972 were intended to protect the attackers."[79] In addition to suggesting the emergence of technology able to achieve a long-time "one bomb, one target" vision for conventional strikes, the 8th Tactical Fighter Wing demonstrated several new capabilities in Vietnam: laser-guided precision strike, electro-optically guided precision strike, and successful night missions. A new family of weapons entered the arsenal, albeit slowly.

Despite the promise of precision munitions, additional funding for new precision strike capabilities did not compete well against other Executive Branch priorities following Vietnam. The Air Force managed to keep a trickle of research and development funding flowing, but did not fight to convince overseers of their relative effectiveness and importance. Culturally, the Air Force still favored its nuclear mission. But as other changes occurred within the Air Force, money was moved from other programs to fund Paveway II development. The improved Paveway had folding wings, allowing more munitions to be carried by aircraft, and a vastly improved guidance package. Paveway II was eventually employed by some thirty nations, with the Air Force ordering some 7,800 conversion kits from Texas Instruments in 1976.[80] That same year the Air Force initiated a Paveway III development program to compensate for significant drawbacks. Most importantly, it had to be dropped from medium altitudes, within enemy air defense weapons ranges. Paveway III was indeed an improvement. Designated the GBU-24, it used "on-board autopilot stabilization so that the bomb could 'cruise' toward the target, a scanning seeker to find the spot of laser light illuminating the target" and had "the ability to be dropped outside the target 'basket' " so it could "maneuver itself inside it."[81]

As support to ground forces became more important, renewed attention was given to airborne surveillance and warning aircraft. The first production version of the E-3 Sentry Air Force Airborne Warning and Control System (AWACS) was delivered in 1975 for testing and evaluation and entered operational service with the 552nd Airborne Warning and Control Wing, Tinker Air Force Base, Oklahoma, in 1977.[82]

During Air Force demonstration flights in Europe, E-3 operators discovered that autobahn traffic was being picked up on radar, fueling interest in a ground moving target indicator (GMTI). Building on ongoing efforts to develop a long-range synthetic aperture radar for the high-altitude TR-1 surveillance aircraft (an updated U2), an Air Force–Defense Advanced Research Project Agency (DARPA) partnership was formed to modify the TR-1 sensor for a GMTI aircraft. This Tactical Air Weapons Direction System (TAWDS), renamed PAVE MOVER, evolved into a project to both identify and track mobile ground targets *and* to attack moving targets with missiles. PAVE MOVER would later become a key component of Assault Breaker.

Soviet SAMs posed additional problems for American defense planners. A number of US aircraft were dual-use, meaning that the same aircraft and pilots responsible for delivering nuclear strikes were also assigned conventional missions including, in some cases, ground support to the Army. In fact, some aircraft were designed for a one-mission battle: the pilot would drop his nuclear payload and then hope to find a friendly airstrip or survive a bailout.

Few believed Air Force commanders would risk their nuclear strike capability to Soviet air defenses to provide tactical ground support. If they failed at the nuclear mission when the time came, their decades-long commitment to nuclear strike would be jeopardized, something that the Air Force simply could not allow. In the end, some posited, the Supreme Allied Commander, Europe (SACEUR) would hold back dual-capable aircraft during the opening phases of war.

Among those making this argument was Alian C. Ethoven, a professor, industry leader, and former senior defense department official. In a 1975 *Foreign Affairs* essay, he contended that "SACEUR will not want to risk losing his nuclear attack aircraft in a conventional war" and "will be strongly tempted to hold them out of the conventional battle and make them, in effect, specialized nuclear forces."[83] Although not all analysts agreed with such arguments, they served to organize debate and discussion about air support to ground forces. Defense analyst and strategist Steven Canby stated the underlying concern: because "a large share of NATO's air forces are still fully committed to a dated Quick Reaction Alert nuclear role," and because "of the demands of air cover and other missions" supporting nuclear missions, "only a fraction of the remainder are available for ground support."[84] By the 1980s, a powerful lobby within the

Air Force succeeded in gaining funding for the F-15 and F-16 fighters, which would dominate the skies for decades to come.

Meanwhile, the Tactical Air Command (TAC) continued to suffer from readiness problems. "On any given day," James Kitfield reports, "half of the planes in TAC's $25 billion inventory were not combat ready because of some malfunction, and 220 aircraft were outright 'hangar queens', unable to fly for at least three weeks for lack of spare parts or maintenance."[85] Ben Lambeth identifies "aircrew proficiency, equipment performance, and concepts of operations" as additional areas of concern.[86]

Training for nonnuclear missions slowly became a priority. Of all the developments that contributed to the thirty-year transformation, training advances remain the most important. This is often underappreciated or overlooked in the story of post-Vietnam American military innovation, a story punctuated with strong narratives of technological change and doctrinal revolution. Nevertheless, training remains the critical enabler of what was dubbed an American revolution in military affairs. Following Vietnam, the Air Force studied combat losses and found that the overwhelming majority of pilots who lost their lives during combat had participated in fewer than ten missions.

Of note was the Air Force's adoption of successful, realistic training techniques previously incorporated into the Navy's successful "Top Gun" program, the US Navy Postgraduate Course in Fighter Weapons established in 1968. In 1978, the Air Force inaugurated its own realistic training program, highlighted by the "Red Flag" exercises held at Nellis Air Force Base. Red Flag exercises pitted new pilots against veterans in highly realistic strike missions in which air defenses were simulated. Instrumented ranges documented the details of every mission to facilitate intensive post-exercise reviews.

Another important development in the post-Vietnam period was the beginning of the Air Force's transition from air power doctrine to thinking in terms of aerospace doctrine. All of the military services pursued, digital, global communications and navigation capabilities, with the Air Force playing the largest role in US military space communications (and the Navy a close second). At mid-decade, the Air Force launched two MIT Lincoln Laboratory experimental satellites (LES-1 and 2). Whereas previous communications satellites had relayed signals between points on the ground, they added a "crosslink" capability enabling both ultra high frequency (UHF) and extremely high frequency (EHF) signals between LES vehicles. Not only did this mean unprecedented communications capability over some three quarters of the earth, but EHF operated at a staggering 38 gigahertz: a much higher transmission rate difficult to jam.[87] Based on a number of advances in satellite communications, in 1979 the Army embarked on a multiyear program to procure hundreds of mobile satellite terminals.

In the early 1980s, and as GPS and other technologies were deployed, Army–Air Force cooperation to leverage emerging space capabilities led to important command and control advances that further linked air and ground power on the conventional battlefield. By the 2000s, distinct views of ground and air operations would give way to discussions of a single "battlespace."

Leveraging the information revolution

During the 1970s, battlefield intelligence began a transformation process that culminated in the 2000s with theater persistent surveillance capabilities and the ability to achieve complete dominance over enemy military forces. A 1974 Army Intelligence Organization and Stationing Study documented many of the problems plaguing battlefield intelligence in the early 1970s. An "indictment of the system that prevailed at the time," the study "found that military intelligence units were not properly organized to support the tactical mission, and, indeed, were in most cases beyond the control of tactical commanders because of their strategic responsibilities."[88] Improving intelligence capabilities was a central focus of technological and operational innovation.

Retired Air Force Lieutenant General James Clapper was the Chief of Air Force Intelligence during the Gulf War, director of the Defense Intelligence Agency (DIA) in the early 1990s and, after several years in industry, returned to government in 2001 as the civilian director of the National Imagery and Mapping Agency (NIMA). NIMA was renamed the National Geospatial-Intelligence Agency (NGA) in November 2003. For his significant restructuring and modernization initiatives at DIA and NGA, he is widely viewed a "change agent" in the intelligence community.

Commissioned in 1963, Clapper served two tours in Southeast Asia during the Vietnam conflict. Comparing his Vietnam experience to his tour as Chief of Air Force Intelligence during Desert Storm in 1991, Lt. Gen. Clapper believes the US had progressed "light years" since Vietnam. In fact, he recalls a sense of surprise, even awe, among some operators regarding the sophistication of intelligence support. Quite simply, posits Clapper, automation, digital communications and intelligence exploitation, and computer-aided command and control available in the 1980s created opportunities for doctrinal change and operational innovation that were impossible just a decade earlier.[89]

Considering early 1970s intelligence support to military operations, Clapper remembers intelligence analysis being important to mission planning but primarily viewed as a support activity. Because of this, Clapper recalled his tours in Southeast Asia as years of "frustration." Intelligence was not recognized as a "main player" in the fight. Bombing accuracies were "awful," targeting accuracies "inconsistent," and exploitation times to feed commanders information from the battlefield "dismal."[90] He recalls that what today would be termed operational intelligence was basically "history" by the time it was exploited and disseminated, useless to pilots on missions over enemy territory.

For example, an American signals intelligence site in Da Nang would often intercept and translate North Vietnamese high frequency manned Morse code communications reporting on US aircraft activity. North Vietnamese units would listen for the morse code messages and use the information about US aircraft activities to coordinate anti-aircraft fire. Unless they were warned, American pilots would not know that the enemy was planning an ambush with anti-aircraft fire. The American signals intelligence units would then attempt to send warning messages after intercepting and translating the North Vietnamese transmissions. The warning messages were sent over sixty-word per minute teletype machines to numerous headquarters in the hope that they would be able to warn their pilots.

Frustration derived from the slow process. Analysts knew that the information they were sending was not getting to pilots in time to alter their attack runs or patrols. They knew that the pilots were not being directed away from the waiting air defense ambushes. Noteworthy here is the fact that signals intelligence was considered by many to be the most effective of the intelligence disciplines in Vietnam, consistently yielding greater insight into enemy activities than human intelligence or imagery intelligence.

Imagery intelligence—then termed photoreconnaissance, also failed to live up to its potential. Not only did triple-canopy jungle conceal activities, there were significant time delays in discovering enemy activities because of processing and interpretation timelines. Even with airborne infrared sensors to detect Vietcong activity at night, film could not be processed and analyzed quickly enough to effect current operations.

In today's parlance, the information was not "actionable." Many of the innovations undertaken in the 1970s aimed to correct this problem, laying the foundation for American military dominance and helping shape visions of future warfighting capabilities that dominated Cold War military thought (e.g. information superiority, dominant battlespace awareness, decision superiority, persistent surveillance, strategic preemption). Two programs contributing to advanced battlefield information capabilities were the CELT and the BETA initiatives. Both were part of the offset strategy that contributed to the evolution of American battlefield information superiority.

CELT, "the first automatic, near real-time system for precision location of communications emitters," was demonstrated during NATO exercises from 1978 through the 1980s.[91] Antecedents included 1960s initiatives to locate North Vietnamese electronic emissions, the Air Force–DARPA Emitter Location System (ELS) project in the mid-1970s, and a Precision Location Strike System. Responding to increased concerns about Soviet conventional forces, ELS was renamed CELT in 1978 and focused on developing "automatic location and classification" of vast numbers of enemy "emitters expected on the European battlefield, with the accuracy required for targeting by standoff weapons."[92] After successfully demonstrating the ability to locate emitters and generate information for targeting, CELT technology and operational concepts would contribute to subsequent systems, including the Army's Guardrail airborne sensor.

Technologies like those being tested in CELT, others being fielded in AWACS, and the array of other ground and airborne sensors either in service already or under development to identify emitters and enemy targets, raised the question of how disparate information sources would be rendered intelligible for decisions.

BETA, created in 1977, responded to concerns about the absence of a "mechanism for correlating and fusing the extensive intelligence information being received from multiple sources."[93] Although there were some fifty studies on the subject, BETA was the first "systemic approach to develop and evaluate" what "correlation and fusion would contribute" by demonstrating "a state-of-the-art, computer-based tactical data facility capable of dealing, in near real-time, with the large amounts of information on the modern battlefield."[94] A 1990 technical assessment of the state of data fusion reported that BETA demonstrated the potential for automated exploitation and targeting "while providing a greater appreciation for the problems associated with data fusion."

Among the lessons applied to later systems included "the need for a disciplined systems engineering approach to future developments," a conclusion that reinforced the emphasis on systems engineering and integration in conventional warfare.[95] An associated development involved thinking about a sensor-to-shooter process derived from complimentary capabilities that were best imagined as interfaces and networks rather than individual platforms.

Related systems included the Joint Tactical Information System (JTIDS) and Joint Tactical Fusion Program (JTFP). JTIDS provided a secure capability to move data around on the battlefield. Less vulnerable to enemy jamming, it worked by spreading data transmissions over different frequencies. JTFP merged a secure communications capability with visualization tools able to represent the fusion of data. Together they aimed to increase situation awareness by providing secure, jam-resistant near-real time information updates to commanders.

Noteworthy is the fact that Vietnam was the first conflict in which military personnel were deployed with a specialty designation "computer specialist." Hundreds operated intelligence and communications equipment. By the end of the 1980s, obtaining computer training and experience in the armed forces would be among the top recruiting draws and a critical readiness area. Recognizing that organizational changes and conceptual innovations are crucial to what this study terms the American RMA, Clapper nonetheless maintains that the advent of the computer information age in all of its dimensions drove innovations in intelligence, weapons development, and doctrine.

These and other efforts, arguably, contributed to a decades-long transformation in which intelligence evolved from an often-marginalized staff or combat support function to a co-equal source of military effectiveness and a leg in a new strategic triad that replaced the Cold War triad of bombers, submarines, and land-based missiles. Soviet conventional readiness and near-real time monitoring of Soviet troop activities ascended on the national strategic intelligence priority list. In the 1980s, the intelligence community increased its exploitation of national intelligence capabilities to provide battlefield situational awareness.

Airborne capabilities also improved dramatically. A combined enterprise of all-weather, day–night sensors became a combat multiplier and improved strategic warning. The United States subsequently perfected capabilities to identify, target, and strike fixed targets and forces whose movements (and probable courses of action) could be anticipated. In the case of the latter, strikes were not necessarily targeting the moving forces as much as known, fixed points along their path.

Less noted, but perhaps historically more notable, was another technological turn. In addition to touting the potential of precision-guided munitions (PGMs) and air defense systems to alter the balance in Europe, 1973 serves as something of a benchmark in electronic warfare. The October War highlighted the value of electronic warfare and influenced the 1976–1977 creation of tactically focused Combat Electronic Warfare and Intelligence (CEWI) battalions at Fort Hood, Texas. Defense analyst Ken Allard views CEWI "as a rather daring innovation that originally incorporated sections for ground surveillance (battlefield radars and ground sensors), electronic warfare, operations security, imagery intelligence, and interrogation."[96] The CEWI had an all-source intelligence production section whose "sole mission was the integration and production of tactical intelligence."[97] CEWIs were an important development in the evolution of the multi-intelligence fusion capabilities that became the cornerstone of American military dominance in later decades.

Electronic warfare was among the six "basic issues" NATO military planners identified in the early 1970s as critical for the modernization of the Alliance's conventional posture. Others were aircraft shelters to protect against surprise attack; anti-armor weapons; war reserve stocks to provide logistics depth to bolster a potential defense; mobile air defense to protect ground forces in light of Soviet frontal aviation advances; and advanced air-delivered munitions to improve interdiction.[98]

Information technology fundamentally changed systems engineering and integration capabilities, which emerged as an important component of military innovation at the theater level, just as they had decades earlier at the strategic level with nuclear command and control. An important legacy of the 1970s was the first 'system-of-systems' meant to enhance situational awareness, an effort that involved linking modified ground-based radars with airborne sensors to enhance the Joint Tactical Information Distribution System (JTIDS). These and other early joint information integration and fusion projects evolved into the common operating picture (COP) and geospatial awareness capabilities central to defense transformation in the 2000s.

During the 1970s, as the information revolution was starting to revolutionize intelligence, Americans received historically unprecedented insights into past wartime and combat intelligence successes. For example, most of the public remained ignorant of the role of intelligence during the Second World War before the 1974 publication of F. W. Winterbotham's *The Ultra Secret*. Ultra was the code name given to intercepted and decrypted communications that had been encoded on an ENIGMA cipher machine, the design for which the British received from Polish operatives before the war. An official ban on referencing Ultra was lifted in 1974, allowing Winterbotham to divulge previously classified information about British cryptography and the breaking of German high command codes.[99] This helped change perceptions about the role of battlefield intelligence.

Fascination with such revelations created a public interest, even appetite, for insights into US intelligence. Meanwhile, public discussions—really criticism—of US intelligence estimates on Soviet military capabilities and the role of the CIA in Vietnam increased awareness of the role of intelligence in foreign policy making. In the 1970s, widespread criticism of American intelligence and highly publicized investigations into abuses of intelligence undermined public confidence in the intelligence community.

Critics charged intelligence analysts with fumbling intelligence reporting on the Soviet threat. In the post-Vietnam political climate public criticism of the CIA was widespread, providing a ready-made platform that notables and intellectuals found useful to leverage. In the mid-1970s, moreover, congressional hearings on rogue CIA covert operations opened the agency to further scrutiny. Leadership turnover did not help. Four different CIA directors served from 1973 through 1976, when George H. Bush assumed the reigns and approved what was known as the Team A–Team B exercise.

The exercise consisted of three teams of nongovernment Soviet experts given access to classified information for the sole purpose of providing an alternate, or competitive, assessment of the same material government analysts used to produce national intelligence estimates, the capstone intelligence assessments produced by the US intelligence community. Teams studied, respectively, Warsaw Pact low-altitude air defenses, the accuracy of Soviet intercontinental ballistic missiles, and Moscow's strategic policies and objectives.

The origins and unfolding of the Team B experiment reflected the national climate. Apparently, its conceptual origins derived from archetypal Cold War strategist, and University of Chicago Professor Albert Wohlstetter's 1974 criticism of CIA national intelligence estimates. He claimed Soviet capabilities were underestimated. Some argued in the late 1970s that "the shocking fact about our intelligence community, with its thousands of able, competent, and dedicated people is, that for 25 years, it has consistently underestimated" the threat.[100] After Reagan's election, many of the Team B members, to turn a metaphor, became the A team in the new administration.

Wohlstetter was also an early advocate for advanced conventional forces, arguing in a 1974 issue of the international affairs journal *Orbis* that increased accuracies made it "possible to use nonnuclear munitions in many circumstances where a desperate hope had formerly been pinned on using small nuclear weapons."[101] Noteworthy for this story of the American RMA is the conjoining of criticism of Carter defense and foreign policy, revelations that Soviet force strength had been underestimated, public discussions of US defense reform, and continued political antipathy to sole reliance on nuclear weapons. Among the issues reformers addressed were doctrinal and technological options to offset Soviet forces. Advanced conventional weapons and new doctrine emerged as a likely, and less expensive, alternative.

Shifts in national security planning

In his acclaimed *How We Got Here: The 1970s,* David Frum argues that changes in the attitude of American citizens toward government affected the context in which leaders in the executive and legislative branches approached defense policy.[102] Early in the decade, Americans seemingly lost faith in the government. No single event or action caused this, despite the importance some assign to Vietnam or Watergate as playing a determining role. Indeed, as Frum argues, "Americans did not lost their faith in institutions because of the Watergate scandal; Watergate became a scandal *because Americans were losing their faith in institutions.*"[103]

Faith was not restored at the end of the decade; criticism of government activities continued. An interesting reversal did occur in defense spending. Whereas the early 1970s brought frustration over the handling of Vietnam and arguments for reduced defense spending in light of detente, the end of the decade brought criticism of defense policy and intelligence for underestimating the threat. At mid-decade "18 percent of Americans said the country was spending 'too little' on defense; in 1978 the number jumped to 29 percent; by 1980, an overwhelming 60 percent worried the country was spending 'too little.' "[104]

What accounts for the change in public opinion? Knowledge of international developments, including increased fears of Soviet aggression, account for part of the change. More importantly, perhaps, was an explosion of domestic debate over Jimmy Carter's foreign policy and defense decisions, a debate aggravated by successive crises, a poor economy, and a sense of national malaise.

Carter assumed office in 1977 with the intent of resurrecting détente to restore stability in Europe, re-focusing American foreign policy on humanitarian concerns, and downplaying the influence of nuclear weapons in world affairs. Promoting human rights resonated with the American, post-Watergate public that wanted national policy to reflect their own sense of ethics and moral responsibility. Rejecting Machiavellian virtues ascribed to statesmen (which prescribed to princes a different moral code than citizens followed), Carter seemingly believed that foreign policy should follow the same principles and codes of conduct expected at home. His initial reluctance to engage abroad militarily reflected personal beliefs as well as the national mood.

Voters had considered Carter's relative inexperience in foreign affairs an asset. Perhaps he would be less likely to entangle the US in protracted conflicts—a concern at mid-decade. Carter did enter office believing his administration could lessen the role military strength played in international affairs and, as Gaddis Smith concludes, "Carter and some of his advisers were readier than any of their predecessors to stare directly at the reality of nuclear weapons" and struggle with "the problems of the nuclear age."[105] Through 1977 and 1978, it seemed the approach might yield progress on several fronts.

Notable achievements included the Camp David accord, normalized relations with China, an un-ratified Strategic Arms Limitations (SALT) II treaty, and a general increase in awareness of human rights issues. On balance, Carter's foreign policy foundered on the critical issue of improving relations with the Soviets. In fact, his repeated attacks against the Soviet human rights record and pursuit of a Washington–Beijing axis of cooperation increased Moscow's recalcitrance.[106] Of course, Soviet leaders were hardly amenable to achieving the full potential of détente, preferring to accept cooperation on select issues relating to Central European stability that reinforced the status quo in Eastern Europe while seeking relative advantage elsewhere.

Carter was also criticized for being overly cautious. Fearing Soviet military aggression and given the relative weakness of flexible response, analysts feared that NATO was vulnerable to "salami tactics" whereby Soviet conventional forces would be used to achieve limited gains in Europe, the Persian Gulf, and in the third world. Conservatives sharply criticized his September 1977 decision to release control of the Panama Canal to the Panamanian government by the end of the century drew sharp criticism from conservatives.

In this context, domestic political forces originating earlier in the decade gained momentum, tapping changes in public opinion to mobilize support. The Coalition for a Democratic Majority (CDM), a group formed in 1972 by conservative Washington senator Henry "Scoop" Jackson, lobbied more aggressively for "peace through strength" and a return to aggressive containment of Soviet expansionism. Members included Senators Daniel Patrick Moynihan, Sam Nunn, and Charles Robb and representative Les Aspin. Academics included Seymour Martin Lipset, Eugene Rostow, Roy Godson, Samuel Huntington, and Richard Pipes. CDM members later formed the core of the "Reagan democrats" supporting increased defense spending in the 1980s.

Aligned with CDM in the cause of promoting increased defense spending were the American Enterprise Institute and the Committee on the Present Danger (CPD). The latter, among the most influential public advocacy groups of the Cold War, borrowed its name from a similar group formed in the 1950s to lobby the Truman administration for increased defense spending. The reconstituted CPD was conceived during a 1976 lunch attended by Nitze, the author of NSC-68. CPD's formal existence was announced just two days after Carter's January 1977 inauguration. Among its members was Ronald Reagan, who would later bring over thirty CPD associates into his administration.

Meanwhile, perceived inconsistencies in Carter Administration policies fueled criticism. A pro-disarmament stance seemed inconsistent with efforts to pressure NATO Allies to increase their military forces. He canceled the B-1 bomber, but initially pursued an advanced radiation (neutron) bomb—later cancelled, approved cruise missiles, and approved accuracy and yield improvements for the Minuteman III missile. The Carter Doctrine and a rapid deployment force seemed out of balance with arguments for nonintervention in the third world. Interest in Moscow's involvement in Middle East peace talks were followed by excluding the Soviets from the Camp Davis peace talks.

Calls for increased US defense spending and for intelligence reform were tightly coupled to arguments for conventional force modernization to reinforce deterrence stability in Europe. President Carter summarized the situation at the 1977 North Atlantic Summit Meeting in London:

> The threat of facing the Alliance has grown steadily in recent years. The Soviet Union has achieved essential strategic nuclear equivalence. Its theater nuclear forces have been strengthened. The Warsaw Pact's conventional forces in Europe emphasize an offensive posture. These forces are much stronger than needed for any defense purposes. Since 1965, new ground and air weapons have been introduced in most categories: self-propelled artillery, movable tactical missiles, mobile air defense guns, armored personnel carriers, tactical aircraft, and tanks. The Pact's build-up continues undiminished.[107]

That same year more than 120,000 Soviet troops deployed into Eastern Europe during a one-week exercise that was part of a normal troop rotation. It was an unexpected deployment that increased fears that an unwarned attack might occur. "The answer," Gaddis found, "in terms both of international events and of what was necessary for [Carter] to retain domestic political support, was to subordinate all other foreign-policy considerations to the rebuilding of military power."[108]

Conventional wisdom held that any crisis outside of Europe would quickly escalate, leading to a global conflict. 1979 was a pivotal year in the evolution of thinking about the Soviet threat to global stability. In September a large Soviet military presence (some 3,000 troops) was discovered in Cuba. A revolution in Iran led to the 4 November occupation of the American embassy in Tehran and the seizure of fifty-three American hostages. Some believed Moscow would aid Tehran. Soviet expansionism in Africa continued. Soviet naval forces became more assertive globally. In December, Moscow sent troops into Afghanistan. Closer to home, the self-proclaimed Marxist Maurice Bishop seized power in Grenada and a revolution occurred in Nicaragua, both setting in motion events that would spur US military action in the 1980s.

One of Carter's last significant foreign policy initiatives involved Persian Gulf security. The Soviet's 1979 invasion of Afghanistan was not part of a Soviet plan to dominate the Gulf, and the terrain did not realistically support a drive into Iran. Nor was Moscow involved in the Iranian revolution. Still, for US defense analysts, the cognitive image of potential Soviet dominance of the region engendered concern that Afghanistan was a prelude to something more ominous.

Apprehension about Soviet military activities in the Gulf steadily increased after the British scaled back their military presence in the late 1960s and early 1970s. The region contained some 75 percent of the world's known oil reserves and supplied a quarter of US oil imports. Western Europe imported roughly 70 percent of its oil from the region; Japan was totally dependant on Persian Gulf oil.

Other developments raised the specter of Soviet mischief in the Gulf. Moscow's ties to the People's Democratic Republic of Yemen evoked the threat of plans to coerce Gulf monarchies, including Saudi Arabia and Oman. Concerns were fueled by a wave of coups, murder plots, and border skirmishes between North and South Yemen in 1978 and 1979. South Yemen formally signed a Treaty of Friendship and Cooperation with the Soviet Union in October 1979.

The situation in Iran contributed to fears of Persian Gulf instability. Ending a thirty-five year US–Iranian relationship, the 1979 Iranian revolution meant the loss of a critical US intelligence facility that monitored, among other things, Soviet missile tests as part of the American effort to monitor and verify SALT. Although Moscow was as surprised as Washington by the revolution, and although the Ayatollah Khomeini marginalized and then crushed pro-Soviet parties, fears of Soviet regional penetration aggravated the sense of American impotence in the Persian Gulf. Soon thereafter, the aircraft carrier USS *Dwight D. Eisenhower* would steam for 251 days into, around, and back from the region, the Navy's longest deployment since the Second World War. The United States also began aiding Saddam Hussein who, in leading Iraq into war with Iran, became an important ally in the struggle to prevent further Soviet inroads into the region.

Soviet expansionism into the Persian Gulf could come by direct invasion, through alignment with a regional client state, or through revolutionary proxies seeking to overthrow a government. Regardless of the mechanism, the United States was unable to block Soviet penetration into the region without a capability for timely and decisive military intervention.

The Carter Doctrine was the name given to the policy of policing the Persian Gulf. It signaled a reversal in the principle underlying the Nixon Doctrine of placing the burden of conventional defense on the nation attacked. The Carter Doctrine also signaled that the United States was once again willing to intervene— no longer moribund by the legacy of Vietnam. It was also an example of containment applied outside of the European theater. Carter subsequently directed the Defense Department to form a military force capable of responding to crises outside of Europe and the Korean Peninsula, at the time the only regional military contingencies that defense planners had previously used to justify modernization.

The Rapid Deployment Force (RDF) was established in late 1979, the year that Saddam Hussein seized power, and was renamed the Rapid Deployment Joint Task Force (RDJTF) in March 1980. Subordinated to the US Readiness Command and headquartered in a converted bunker at MacDill Air Force Base in Tampa, Florida, the RDJTF was chartered to prepare plans, engage in joint training activities, conduct exercises, and to be ready for deployments into areas where US interests were threatened. Within a year of its creation, the RDJTF would complete some ten training exercises.

Increased attention to Persian Gulf security focused defense planning on understanding the requirements for moving military forces into the region and sustaining combat power. Because this required deploying troops from Europe, Allies decided in May 1980 to accept the idea of redeploying US forces protecting NATO from Europe to the Middle East during a crisis. Of note to students of US Middle East policy, this was the first NATO agreement to support what post-Cold War NATO planners term "out of area operations."

Although it would be more than a decade before the RDF was able to deploy significant combat power "rapidly" to regions other than Europe, the rapid deployment concept was an early manifestation of what current military thinkers term rapid decisive action, a concept that reinforces the image of preemptive strikes. In 1979, Brzezinski suggested that the RDF would be used preemptively "in those parts of the world where our vital interests might be engaged and where there are no permanently stationed American forces."[109] "It is not necessary," Secretary of Defense Harold Brown similarly stated, "for us to await the firing of the first shot or the prior arrival of hostile forces."[110]

Chapter conclusion

When William Perry and Harold Brown first articulated the offset strategy in the late 1970s it was not clear what effect it would have on the East-West military balance. It was certainly not envisioned that the offset strategy would mature into a new, dominant form of strategic warfare in the early 2000s. None imagined the marginalization of nuclear doctrine in mainstream defense planning some two decades later after the Warsaw Pact dissolved without a shot fired. Moreover, no mainstream analyst envisioned the emergence of a national military strategy dominated by information superiority, precision weapons, and what by all accounts was an offensive doctrine.

On the other hand, they *did* conjecture that proposed conventional systems, including organizational adaptations redressing long-standing Army–Air Force issues, might make it possible to operate on a "porous battlefield," one without massed forces or linear battle lines, in a manner that did in fact render tactical nuclear weapons "redundant."[111] "One common characteristic" of new technology, M. W. Hoag surmised in the early 1970s, "might be wide dispersal of small ground combat units, to replace the contradiction between massing for conventional combat and the need to disperse for a 'nuclear-scared' configurations."[112] If accompanied by "improved posture and tactics, and improved night and low-visibility target acquisition and guidance systems," other analysts argued, "the contribution of these weapons systems to stopping a Pact ground offensive could be decisive."[113]

Research and development pursuant to the offset strategy encouraged the aligning of new capabilities to conventional missions. Early applications for distributed information technology included sophisticated, collaborative, realistic, and unparalleled training and simulation initiatives. New thinking emerged about the role of conventional forces in theater and strategic-level planning. An explosion in conventional deterrence literature is perhaps the most fundamental artifact of this thinking.

Technology was only a part of the story. Military leaders embraced innovation in the late 1970s. Preparing to actually fight—and perhaps win—a conventional conflict impelled new thinking about deep strike. Operational challenges posed by the threat of a Soviet armored attack engendered a new approach to the conventional defense of Europe. Planners reconsidered maneuver and revamped the overall approach to interdiction. Because it seemed unlikely that NATO would make a nuclear release decision quickly, and because Soviet planning aimed to flood NATO's front with breakthrough-oriented armored attacks, Army units on the front line faced quick death or capture. Doctrine and new operational concepts provided the essential integration medium through which new technology was brought into the larger military innovation process.

An overhaul of Army training significantly improved battlefield performance. The National Training Center (NTC) at Fort Irwin, California was conceived to provide a level of realism dearly needed to prepare for the defense of Europe. NTC remains the world's most sophisticated armored warfare battle laboratory for experimenting with new concepts and weapons and for training unit commanders to think creatively. NTC provided a testing ground for the "big five" conceived in the 1970s as the core weapons systems of the revitalized Army. New systems included the M1 Abrams main battle tank, M2 Bradley armored infantry fighting vehicle, AH-64 Apache combat helicopter, UH-60 Blackhawk utility helicopter, and the Patriot air defense missile system. Linking them together were a new arsenal of information and command and control systems that multiplied the baseline capabilities of individual weapons systems.

For Vice Admiral Bill Owens (retired), the capabilities labeled revolutionary in the early 1990s derived from operational approaches and systems "engineered and acquired in the late 1970s through the late 1980s" that made victory in the 1991 Gulf War "inevitable and our historically small loss of life probable."[114] Former Assistant Secretary of Defense for International Security Policy (1993–1996) Ashton Carter adds that after the Offset strategy's precepts were "dramatically demonstrated during Operation Desert Storm" they became "key to Washington's way of waging war."[115] Chapters 5 and 6 explore this assertion, finding much to support the importance of the Offset strategy in the evolution of the new American way of war.

After the offset strategy jelled as an overarching vision for US defense planning in the late 1970s, the initial steps to implement it fell to technologists. DARPA played a pivotal role. Working closely with Service counterparts and drawing on studies like the above-mentioned 1976 Defense Science Board study, the research and development community benefited from insights into crucial operational requirements. Intelligence analysts informed the process from the beginning.

Intelligence, surveillance, and reconnaissance received increased attention among national security scholars in the 1970s, in part due to public debate about the validity of US intelligence estimates, Congressional committees and investigations into intelligence operations, and revelations about Ultra. Knowledge of increased Soviet nuclear and conventional capabilities, coupled with Soviet moves into the third world, renewed concerns about surprise attack. Defense planners subsequently called for improved national and theater strategic warning capabilities.

Concerns also increased that any crisis would lead to uncontrolled escalation and nuclear war. With US cruise missiles and Pershing II missiles deployed in Europe, some European leaders compared the strained East–West relationship and reciprocal military buildups to the pre-First World War environment. Misinterpretation could easily spark war, something that added a sense of impending crisis to an already heated domestic policy debate.

Heightened fear of war led to the first national-level security planning document on command, control, communications, and intelligence (C3I) in November 1979. Among the outcomes were national objectives for C3I, including a telecommunications capability to preserve command of nuclear forces, and research and development activities that led to the Internet's creation. A new continuity in government operations directive followed a year later. March 1980 brought the first updated national guidance on mobilization in nearly twenty years. Carter signed Presidential Directive (PD) 59, shifting US nuclear targeting policy to new strategic military targets and mandating that the US be able to fight a protracted nuclear war. Adhering to PD 59 required further C3I advances, more flexible targeting, and new nuclear weapons.

Integral to the Offset strategy were changes in defense acquisition processes, an area that warrants additional study. Brown acted to improve the prevailing "system of acquisition that basically was conceived in reaction to our failures" and to improve, relative to the Soviets, the US ability "to translate available—and roughly comparable—technology and productive capability into the most effective military posture."[116] To achieve this, a formal process for developing and coordinating "mission needs" was developed, with a Mission Element Need Statement required in addition to other documents. Over time, as military forces were increasingly integrated with a joint warfare framework, mission needs evolved into mission capability packages.

After an initial emphasis on nuclear capabilities in the early 1970s, US defense strategy slowly shifted to conventional warfighting requirements. By the end of the decade the shift would be quite pronounced compared to the nuclear-centric defense planning discussions of the late 1960s. Flexible response nonetheless remained the dominant undercurrent in strategic planning discussions, but now there was a concerted effort to fund new nonnuclear weapons systems.

The change began with initiatives undertaken at the end of the decade, notably President Carter's reprogramming of defense dollars to improve European defense readiness. European manpower levels rose from some 300,000 troops in 1975 to 330,000 at the end of the decade. Troop strength increases were bolstered by significant research and development activities to modernize forces, including a new family of weapon systems. Decisions about defense research and military innovation made during the 1970s established momentum that was fundamental to the Reagan defense buildup.

Foreign policy and defense issues dominated the 1980 presidential election. Debate over the severity of the Soviet–US military balance continued. Both the domestic political situation and worse-case analysis of the balance of power warranted action. "By this time," one analyst concludes, "Soviet factories were busy spitting out an average of five fighter planes, eight tanks, eight artillery pieces, and at least one ICBM every day."[117] Writing in *Aviation Week and Space Technology*, retired Major General George Keegan, Jr, former chief of Air Force intelligence, asserted that the United States "lacks the firepower, lacks the accuracy, and lacks the yields to overcome the enormous advantage in terms of neutralizing our retaliatory punch which the Soviets have engineered for themselves at great cost."[118] As Secretary of Defense Donald Rumsfeld had earlier proclaimed in his Fiscal Year 1978 Posture Statement to Congress, "the burden of deterrence" had "fallen on the conventional forces."[119]

5 Expanding missions, new operational capabilities

A renaissance in American military thought occurred in the 1980s. The doctrinal innovations that shaped post-Vietnam military renewal coincided with a training revolution that emphasized fighting outnumbered and winning the opening battles of a conflict in Europe, the Persian Gulf, or other region. Greater dynamism in all combat operations reinforced a belief that temporal and spatial dominance, audacity, and flexibility were key elements of the approach to winning the opening battles of a war. Speed in decision making and operations drove thinking about operational art, which became a core area of leadership training. Battlefield command and control emphasized simultaneous operations throughout the depth and breadth of the battlefield. An affinity for automation deepened on both sides of the Iron Curtain as battlefield intelligence was more tightly coupled to strike systems. Decision timelines previously measured in hours were compressed to minutes. Meanwhile, the volume and precision of indirect fires available to commanders increased. The overall capabilities of conventional forces increased significantly.

The 1970s struggle to redefine deterrence strategy and reclaim deterrence stability was motivated in part by the simultaneous emergence of NATO–Warsaw Pact nuclear parity and the need to offset Soviet conventional forces. Political pressure to reduce reliance on nuclear weapons complicated NATO war planning. Defense analysts, moreover, questioned the viability of flexible response strategy and revisited the need for a more diverse range of defense and retaliation options to counter Soviet military power. Deterrence strategy subsequently underwent a reformation as theorists and strategists expanded their emphasis to address the possibility of conventional deterrence.

Retired Lieutenant General (LTG) James C. King participated in the Army's post-Vietnam renewal. King was successively the Director of Intelligence for the European Command, the Director of Intelligence on the Joint Staff, and the Director of the National Imagery and Mapping Agency (renamed the National Geospatial-Intelligence Agency in 2003). After combat tours in Southeast Asia, King played an important role in the Army's development of a new family of tactical intelligence and theater targeting systems, including the overhaul of electronic warfare capabilities. He agrees that a period of innovation was ushered in at the end of 1970s, leading to significant capability increases in the 1980s.

King argues that Army commanders had adopted a different mindset by the 1980s. He recalls being the operations officer for an intelligence battalion in the Army's 7th Corps deployed with the 3rd infantry division for a massive Return of Forces to Germany (Reforger) exercise in 1977. Senior Army leaders were becoming more open to innovation. A sense of urgency was evident as leaders attempted to transform Army training and doctrine as they sought to rebuild American military superiority.

Recognizing the need for new thinking, leaders were highly receptive to technological and organizational innovation. If someone had an idea or tech-nology that had proven itself in testing and evaluation, the Army's European Command, the primary Army force provider to NATO, would try it. The shift in mindset, King recalls, reflected readiness to integrate new concepts, operational approaches, and weapons systems into Reforger even if doing so disrupted the exercises. It was during this period that experiments and technology demonstra-tions confirmed the value of applying information technology.

King remembers the strategic situation—having to delay Soviet conventional forces for three weeks—as a prime motivator of innovation initiatives.[1] Operational necessity encouraged innovation throughout the Army as commanders struggled with the realization that their units were not trained, that subordinate commanders were unprepared mentally for combat, and that their weapons sys-tems and communication systems were quickly becoming outdated. Assessments of combat readiness and unit effectiveness continued to suggest that American forces would likely fail to succeed in their mission of delaying the onslaught of the Red Army's armor, helicopters, and artillery.

Changes in the strategic context spurred changes in the calculus of strategic effec-tiveness, which in turn altered definitions of military effectiveness. New roles and missions emerged for US military takes both inside and outside of the European the-ater. Visionary leaders seized opportunities to pursue military innovation, including refocusing doctrine and training at the operational level of war. Meanwhile, research and development activities begun in the 1970s were starting to demonstrate their promise to alter the strategic balance in Europe. New nonnuclear weapons systems and operational concepts were slowly integrated into the armed forces. Military officers subsequently rediscovered maneuver warfare in the 1980s and reawakened the offense spirit historically attributed to American military culture.

In terms of innovation theory discussed in Chapter 2, by the end of the 1980s, it appeared that the core organizational competencies of Cold War armed forces were being challenged by the appearance of doctrinal, technological, and operational innovations. Willingness to try new capabilities partly derived from the above mentioned perceptions that the armed forces should vigorously pursue precision munitions, advanced automation, and other capabilities emerging from the digital age and the computer revolution. Military planners, moreover, realized that information technology offered solutions to time and distance challenges. Spatial and temporal distance no longer determined communication parameters, a critical factor for a military contemplating long-range precision strike and dynamic re-targeting of nuclear missiles.

By the end of the 1980s, nuclear strategy no longer dominated national security and defense strategy, opening intellectual space for new ways of thinking about and planning for war. New concepts, visions, and operational approaches from the information technology domain migrated to fill this intellectual space.

The conventional turn in American military strategy, including the advent of global precision strike—a centerpiece of American grand strand strategy in the 2000s, was conceived in the late 1970s but did not become established until the mid-1980s. This chapter addresses the key part of this story, focusing on developments that gained prominence as the Cold War ended.

New roles and missions were added to the calculus of military effectiveness as the Cold War ended with the November 1989 fall of the Berlin Wall, the 1990 decision to reunify Germany, and the July 1991 dissolution of the Warsaw Pact. In the 1990s and 2000s, there was a noticeable shift in how resources were mobilized to achieve grand strategy, which Barry Posen defines in *The Sources of Military Doctrine* as "a political-military, means-ends chain, a state's theory about how it can best 'cause' security for itself."[2] This shift, which includes the ascent of nonnuclear global strike as a pillar of American national security strategy, is a legacy of the Cold War.

The Reagan buildup and shifts in national security planning

Ronald Reagan was a staunch anti-communist who campaigned against Jimmy Carter on a platform promising to restore American power overseas and pledging to restore confidence at home. He promoted a vision that included reducing the threat of nuclear war, ending Moscow's occupation of Eastern Europe, and restoring America's belief in itself as world leader. His two terms in office (1980–1988) witnessed the largest and most rapid peacetime military expansion in the history of the United States. Of course, the buildup began during the last years of the previous administration. Carter's January 1980 State of Union address announced the renewal in registration for the draft and a 5 percent increase in defense spending, from 165 billion in fiscal year 1981 to 265 billion (constant dollars) in 1985.

Reagan administration insider and economist Martin Anderson distilled Ronald Reagan's personal views and policy initiatives into a set of precepts he believes constituted Reagan's "grand strategy." Although Reagan never formally articulated or documented them, Anderson believes they nonetheless guided his thinking. Reagan believed that any nuclear war between the United States and the Soviet Union would have unacceptable, devastating consequences. He found a national military strategy based on mutually assured destruction to be morally repugnant. He was committed to reducing nuclear weapons, not merely freezing or otherwise limiting certain classes of weapons. Expressing deep skepticism about complex arms control processes, he preferred direct negotiations with leaders.

For Reagan, the Soviet Union was "an implacable foe" that controlled "an evil empire" that America had a duty to confront. Since at least 1978, Reagan had been publicly stating that Moscow was "an evil influence" working across the globe in "far-flung trouble spots" to further its "imperialistic ambitions" by "stirring a witches' brew."[3] Most Americans old enough to remember Reagan recall his March 1983 speech condemning Soviet leaders for being a focus of evil, his famous "evil empire" speech.

Reagan expressed a fundamental belief that America's economic and moral superiority could be leveraged to "upgrade the power and scope" of American military capabilities and gain acceptance for American engagement abroad. He hoped that "the Soviets would not be able to keep pace" with American economic and military expansion.[4]

Reagan was not alone in his opposition to Soviet policies. Margaret Thatcher became the Prime Minister of Britain in 1979—some twenty months before Reagan's January 1981 inauguration. A key ally in fighting communism, she had earned the nickname "Iron Lady." One day after Reagan's evil empire speech, the British leased the island of Diego Garcia to the United States to serve as a base for strategic bombers and warships. Other NATO leaders supported Reagan's policies. Francois Mitterrand became the President of France toward the end of Reagan's first year in office and for a period displayed uncharacteristic French support for NATO. Domestic political dynamics helped soften Mitterrand's leftist leanings. Another conservative, Helmut Kohl, became Chancellor of West Germany in 1983. He approved the deployment of Pershing Missiles to US military bases in Germany that November.

Western leadership grew more conservative; Soviet leadership initially evolved in the opposite direction. Yuri Andropov died in February 1982 at the age of eighty-four. His replacement, tuberculosis-stricken Konstantin Chernenko, held power for fourteen months before his March 10, 1985 death. He was the last of the old guard communists bent on preserving a global struggle against capitalism.

Changes in NATO and Soviet leadership occurred against a backdrop of increased Western fears of Soviet military power. Increased militarism in Soviet foreign policy seemed more than mere rhetoric in light of Soviet naval activity and additional posting of Soviet military advisors to client states. New SS-20 mobile missiles further threatened deterrence stability, spurring interest in controlling warhead and missile technology proliferation. The discovery of Soviet troops in Cuba heightened fears that the Kremlin would use Cuban troops to subvert Caribbean governments and gain a communist foothold in the Western Hemisphere. Large Warsaw Pact military exercises and increased nuclear alert levels suggested a more aggressive and effective Soviet military.

This study does not question whether the Reagan administration indeed entered office with a coherent, policy-focused grand strategy. Reagan himself was widely criticized for being ill equipped to make complicated foreign policy decisions and few in his inner circle were considered veteran statesmen. On many issues, however, he appears to have been remarkably prescient.

Like Carter before him, Reagan sought to impel a Soviet turn toward openness and an improvement in the Soviet human rights record. How Reagan set out to affect such changes in Soviet policy marked a clear break with his predecessor. Important to this study were defense and national security policy initiatives related to the Reagan administration's quest to overcome perceived Soviet military advantages and push the Soviet empire toward exhaustion.

In 1981, the CIA estimated that Soviet annual military spending was roughly double the US spending in real terms. A year later Soviet factories were producing "1,300 new fighters a year, about three to four times the fighter replacement rate of the U.S. Air Force" and an average of "a squadron a week and a wing a month.[5] Defense analyst Ralph Sanders concluded that Soviet industry was turning out "about three times as many tanks and armored vehicles, twice as many tactical combat aircraft and military helicopters" by 1985.[6] For those concerned with Soviet naval developments, which threatened sea-borne reinforcement of NATO and became an important consideration for Persian Gulf contingencies, the Soviets also produced "four times as many attack submarines" and nearly "the same quantity of surface combatants."[7]

The conventional military threat to European stability seemed particularly grave in light of the 1981 Polish crisis. Soviet military units practiced radio silence, preventing US signals intelligence assets from reporting on signs of troop movement. Shorter winter days meant less daylight to take satellite imagery of troop dispositions or changes in their mobilization status. The weather further degraded the satellite imagery because electro-optical imagery satellites cannot penetrate through clouds. A number of factors left US commanders with "no idea" of where significant numbers of Soviet troops were. Divisions moved. Some movements went undetected for days, even a week.

Robert Schulenberg, then director of the US Army's indications and warning office within the United States European Command, was among a handful of analysts with access to highly classified, all-source intelligence. He recalls that the 1981 crisis served as a catalyst for modernizing strategic and theater intelligence collection systems as well as enhancing information sharing among analysts.[8] In the aftermath of the crisis, assessment teams visited units responsible for monitoring Soviet troop movements and providing early warning of changes in Soviet mobilization potential. Additional theater day/night, all weather, near-real time intelligence collection capabilities were developed to prevent another gap in warning.

Back in Washington, Reagan put in motion events that would redefine the overall strategic framework guiding American security policy. The President signed National Security Study Directive 1 (NSSD-1) on February 5, 1982, signaling that the traditional concepts such as containment, mutual assured destruction, and détente—the status quo approach to the East–West relationship—were no longer acceptable. The Cold War was no longer considered a "permanent condition"; NSSD-1 aimed to establish steps the United States could take to end the Cold War "on a basis acceptable to American values."[9] National Security Council staff member Thomas Reed, later Secretary of the Air Force, chaired NSSD-1's interagency review of "U.S. national security objectives and the impact of Soviet power and behavior on those objectives."[10]

NSSD-1 provided background for three seminal National Security Decision Directives that "laid out the institutional arrangements and component parts needed to push the Soviet Union to the wall" and "bring the struggle to an end on America's terms."[11] NSDD-32, signed on May 5, 1982 outlined steps to "neutralize efforts of the USSR to increase its influence."[12] Reed states that NSDD-32's "bottom line" was bringing about "the dissolution of the Soviet empire" using an integrated strategy consisting of "economic, political (at times to include covert action), diplomatic, information (both the promotion of unfettered communication and the use of propaganda), and military (to include arms control) ends."[13]

While not comprehensive, NSDD-32 provided a departure point for more specific directives and a framework to coordinate activities across the national security community. On November 29 NSDD-66 was approved. It "codified a stiffened plan for economic relations with the Soviets."[14] Reagan's "full court press" to bankrupt the Soviet economy included "denying [them] critical resources, hard currency earnings from oil and natural gas exports, and access to Western high technology."[15]

A more ambitious approach was outlined in NSDD-75. Reed describes it as providing "the blueprint for the endgame" of the Cold War by calling for additional measures to reverse Soviet expansionism, promoting pluralism within the USSR, and engaging the new generation of Soviet leadership.[16] Rather than seeking accommodation with Moscow or a rekindling of détente, Reagan sought to compel change in Soviet behavior through increased military spending, innovative economic initiatives and sanctions, and a program to question the very legitimacy of the Soviet presence in Eastern Europe.

Signed on January 17, 1983, NSDD-75 specifically aimed to raise the "costs of empire," to roll back Soviet expansionism outside of Europe, and to undermine the power and influence of the Soviet elites that controlled all aspects of life in the USSR.[17] Additional radio and television broadcasts were part of the plan, as were highly classified activities to isolate Poland from the Warsaw Pact and to aid Polish writers, dissidents, and opposition leaders fight the tyranny of Soviet occupation. Addressing the existence of the "evil" Soviet empire later that March, Reagan would comment privately to friends, "It's time to close it down."[18]

Meanwhile, political and economic resources were mobilized in support of military innovation. Reagan sought some eight billion dollars in supplemental defense spending to augment the already increased fiscal year 1981 defense budget. Because republicans controlled the Senate, Reagan had a political base to secure Congressional support for additional defense spending. He subsequently asked for and received an eighteen billion dollar supplemental for fiscal year 1982. Building on Carter's planned defense increase, military spending more than doubled between 1981 and 1986. A large share went to nonnuclear forces, although spending on nuclear forces also increased. In addition to conventional forces, Reagan restored the B-1 bomber program, expedited deployment of the MX and Trident missiles, and increased the intelligence budget. Strategic missile defense technology and advanced precision munitions were among the administration's legacies.

Throughout Reagan's two terms it was often the civilian leadership of the Defense Department or Congressional leaders that pushed innovation. Defense department civilians mandated improvements in conventional early warning, championed funding to increase the readiness of European ground and air units, funded some deep strike conventional weapons systems that were not popular within the military services, and campaigned for rapidly deployable forces able to check Soviet (or Soviet-backed) aggression outside of Europe. Seeking to accelerate the fielding of new capabilities, Congress directed the formation of an Office of Conventional Initiatives within the Office of the Secretary of Defense to integrate programs, expedite innovation, and to deliver new capabilities to operational units more quickly. In the mid-1980s, civilians enacted Packard Commission recommendations to change the acquisition process and expedite weapons procurement processes. Civilians also rescued the Strategic Computing Initiative and other computer programs targeted for elimination as a cost-saving measure.

In 1986, the Goldwater–Nichols Act mandated a new framework for the planning and conduct of joint operations, a new training and development program that included mandatory joint assignments for senior officers, joint program development, and other enablers of the "joint warfighting capability" that came to define the American way of war. The Act was passed over the objections of senior military leaders.

Reagan pressed for increased defense spending and a more visible military presence abroad at a time when he enjoyed widespread public approval. He used his approval ratings as leverage to sustain Congressional support. It is important to note that Reagan rode the shifting tide of public opinion as much as he shaped it. In the mid-1970s public opinion polls generally indicated that the majority of Americans did not favor military engagement abroad or increased military spending. Indeed, some surveys found that Americans did not favor the use of troops to defend any country other than Canada. One survey recorded a narrow majority actually opposed to American troops defending Western Europe. By the end of the 1970s, the tide was already turning, with a majority willing to send troops to fight not only in Europe but to defend the Persian Gulf, Japan, and even Pakistan.

Additional military engagement abroad was a cornerstone of Reagan's attempt to reverse Soviet expansion. Defense planning guidance adapted Carter's one-and-a-half war planning scenario and the Carter Doctrine. The United States needed to respond to Soviet aggression in all regions simultaneously, preventing Moscow from consolidating in one region as the Soviet empire was challenged on all fronts. Reagan Defense Secretary Casper Weinberger's planning guidance described "a worldwide war" plan that called for "concurrent reinforcement of Europe, deployments to south-west Asia and the Pacific, and support for other areas."[19] Media reporting of the fiscal year 1982 defense guidance provided insight into just how dramatically planning had changed from the Carter administration. The 1982 guidance discussed the possibility of actually "winning" a limited but protracted nuclear war involving exchanges over several months.

In light of worsening US–Soviet relations, Reagan saber rattling and talk of winning nuclear wars spurred a political backlash. Political pressure increased for reduced reliance on nuclear weapons. In 1983, an influential group of Catholic Bishops condemned nuclear war, although they did not go as far as condemning the existence of the weapons themselves. Convinced that the same scientists that invented nuclear weapons could improvise a morally superior defense against them, Reagan announced what later became the Strategic Defense Initiative.

Nuclear weapons opponents changed their tactics during the 1980s. Instead of arguing against war in general or the moral legitimacy of nuclear weapons, they engaged in debates about the utility of nuclear weapons and the consequences of using them. The post post-Vietnam, post-Watergate context, furthermore, led to more open questioning of government positions and policies, especially on matters related to defense. Widely publicized studies argued that even a relatively small nuclear exchange—a limited nuclear war—would result in environmental catastrophe, systemic economic collapse, and prolonged political disruption. As few as 240 Soviet warheads targeting American liquid-fuel sites would destroy the US economy and kill 60 percent of the American population within three years. Some even argued that a nuclear winter would occur, putting an end to human civilization.

Paul Herman served as a senior defense analyst and later headed the Alternate Futures Project. He describes the anti-nuclear sentiment among policy makers as an "internalized" and "widespread revulsion" to "weapons whose main property is to kill or maim people (versus destroy their armaments)."[20] Underlying this revulsion was growing anxiety concerning the use of nuclear weapons in general, a cognitive tension wrought from the dissonance of simultaneously relying on these weapons for peace and security while morally rejecting their existence altogether. Reagan reportedly felt this same dissonance, leading to his steadfast determination to end the Cold War. Opposition to nuclear weapons increased throughout the decade. Criticism focused on plans to deploy W79 and W70–3 "enhanced radiation" warheads designed to be fired from tactical weapon systems in Europe. These so-called neutron bombs were designed to blunt Soviet armor by boosting the radiation output of a smaller-yield weapon. This would deliver an incapacitating radiation dose to enemy troops without the blast, heat, and other effects a larger yield warhead would produce, thereby causing less damage and fallout. Their lower yield combined with greater operational utility sparked international controversy because they appeared to lower the threshold for nuclear weapons use in a future conflict; at the very least they might spark a new arms race to develop tactical weapons. Bowing to political pressure, Congress banned any for funding for or fielding of enhanced radiation warheads.

Nuclear weapons were not the only focus of criticism. Observers of the system of nuclear deterrence, including launch-on warning retaliation plans, questioned whether the combination of algorithm-based early warning systems and rapid launch processes might lead to inadvertent nuclear war. Fears seemed warranted in 1983 when a malfunctioning Soviet nuclear launch warning system erroneously reported that NATO was attacking the Soviet Union. According to one account, "because the duty officer of the day came from the algorithm department and could sense that the alert was inauthentic" the report "was not relayed to the Politburo."[21] If it had not been for the officer recognizing the alert as an anomaly, some contend the Kremlin would have been notified that an attack was underway, leading to unknown consequences. Derek Leebaert relates that a classified 1989 US assessment of the 1983 "war scare" is "terrifying."[22]

All of this reinforced the quest for alternatives to nuclear weapons in NATO defense planning and to shift the focus of American military strategy from nuclear targeting. Arguments for "raising the nuclear threshold" were revisited, including those underscoring the doctrines of Follow-On Forces Attack, Deep Strike, and AirLand Battle. The Commission on Integrated Long-Term Strategy provides one of the historical benchmarks in this process. It concluded in 1988 that "the precision associated with the new technologies will enable us to use conventional weapons for many of the missions assigned to nuclear weapons," and that, "as accuracy improves, the nuclear yield needed to destroy hardened military targets also drops dramatically, to the point where conventional warheads could do the job."[23] Similar arguments engulfed defense panning discussions in the early 1990s as US national security strategy was overhauled and the emphasis on nuclear weapons decreased.

Additional emphasis was placed on automation for both nuclear and nonnuclear command and control. Rapid response remained a key tenet of defense planning, despite arguments against launch on warning. The capabilities to launch a weapon quickly remained a central component of deterrence. Planners believed that automating warning-to-launch systems would shorten decision making processes, make them more efficient in terms of simplifying the information that would be communicated to the launch officers that remained the critical step in the process. Modernization of nuclear command and control became a strategic imperative. Cutting decision making timelines drove the application of new technology, including research into the distributed communications systems that evolved into the Internet. Expedited, secure, and survivable command and control enhanced deterrence through mission assurance: even if surprised and already under attack, NATO forces would be able to retaliate. At the same time, Reagan increased funding for strategic missile defense.

Concurrent with the modernization of nuclear command and control, defense analysts inside and outside of the government continued to emphasize the strategic imperative to modernize conventional forces. The Defense Science Board (DSB) and other defense advisory groups continued to focus attention on the potential for emerging technology combined with new operational approaches to offset Soviet conventional forces and to lower the threshold for nuclear use in a future conflict.[24]

Arguments for conventional modernization highlighted the need for additional defense modernization funds, leading to renewed attacks on Carter defense policies. Because Moscow "spent in real terms some $185 billion more on military [research and development] R&D" between 1975 and 1985, US defense analysts concluded that Soviet forces were "able to deploy one-and-one half or two generations of equipment, while the United States has been able to deploy equipment one generation old."[25] Soviet defense spending as a share of Gross National Product increased from 12 percent in 1965 to 17–25 percent in 1985, depending on the methodology used for calculation.[26]

As the American defense buildup gathered momentum and strategists reconsidered the role of nuclear weapons, the underlying structure of Cold War strategic relations changed dramatically. Mikhail Gorbachev was elected General Secretary of the Communist Party in March 1985 and in July Eduard Shevardnadze succeeded Andrei Gromyko as Foreign Minister. The implications of Gorbachev and Shevardnadze's pursuit of "new thinking" in Soviet foreign affairs were demonstrated at the November 1985 Geneva Summit, a turning point in US–Soviet relations. It was the first Soviet–American summit in nearly a decade. Reagan and Gorbachev agreed to accelerate disarmament talks. The United States welcomed a Soviet consulate in New York and opened an American consulate in Kiev. At the end of the summit Gorbachev uttered a sentence during a ninety-minute press conference that, in hindsight, aptly describes the historical importance of his ascension to power: "The world has become a safer place."

Gorbachev would later publicly address one of the thorniest issues in the US–Soviet relationship a year later: Soviet occupation of Afghanistan. Gorbachev's opening speech to the February 27, 1986 27th Party Congress referred to Brezhnev's tenure as "years of stagnation" and called Afghanistan a "bleeding wound." At least parts of the world seemed safer in terms of Cold War conflagration. In the vacuum left by Soviet troop withdrawal simmered an even more incipient threat to world peace, the rise of global jihadist extremists that the United States would wage war on with the weapons, doctrine, and overall approach to regional conflict that were developed in the 1980s.

A second Reagan–Gorbachev summit in Reykjavik, Iceland on October 10, 1986 resulted in broad agreement on arms control issues. Former Secretary of State George Shultz and British Prime Minister Margaret Thatcher considered the summit a "turning point" in the Cold War's end; former National Security Advisor Zbigniew Brzezinski concludes that "it was at Reykjavik that the Cold War was won."[27]

The Cold War's demise proceeded apace. In December 1987, Gorbachev visited Washington, DC and signed the Intermediate Nuclear Forces agreement. Two years later Gorbachev unilaterally withdrew 500 non-strategic nuclear weapons from Eastern Europe. In February 1990, NATO and the Warsaw Pact leaders agreed to German unification, which occurred that October. A month later the Treaty on Conventional Forces in Europe was signed, stipulating that the Soviet Union would reduce its conventional arsenal by nearly 50,000 tanks, armored vehicles, artillery pieces, and combat aircraft. During this time, the United States ended Looking Glass command and control flights. Looking Glass aircraft had operated continuously since 1961 to ensure that the United States could command its global nuclear forces in the event of a surprise attack on American military installations. The Warsaw Pact dissolved in July 1991.

By the end of the Reagan years, as the American approach to military modernization was qualitatively different from past practices, primarily due to the advent of the microchip-based information revolution and the ascent of "systems of systems" thinking among defense reformers and military strategists.

Time-dominance, interdiction, and battlefield integration

Major General Robert Scales, Jr. (retired) commanded a field artillery battery with the 101st Airborne division in Vietnam. "After Vietnam," Scales concludes, "the fortuitous development of revolutionary information and precision technologies gave the US military a means to overcome past inefficiencies," giving rise to a "new American way of war" in the 1980s.[28] Its impetus included the "premise that technology could kill the enemy faster than the enemy could find the means to offset this overwhelming advantage in formation and precision."[29] By the end of the 1980s, information systems and decision technology developments were maturing toward the objective of affording commanders unparalleled situational awareness.

Although many problems remained, including cross-service interoperability challenges, important advances in intelligence, surveillance, and reconnaissance transpired; a revolution occurred in America's overall command and control capabilities. Weapons systems integrated and enabled by information technology led to operations being more lethal over greater distances with far fewer forces. Institute for Defense Analyses researchers Richard Van Atta and Michael Lippitz concluded that, "the ability to exercise military control [shifted] from forces with the best or the most individual weapon systems toward forces with better information and greater ability to plan, coordinate, and accurately attack."[30]

Other developments in the 1980s set the stage for advances in intelligence support to military operations. This included more timely and precise information for target identification and location. Better information for precision strike opened additional routes to develop smart weapons with computerized guidance and navigation systems. The automated information fusion and decision tools developed as part of the offset strategy and Assault Breaker program spurred additional efforts to expedite command and control of battlefield strikes and to integrate additional systems into a larger, more capable command and control network. How planners thought about force modernization shifted, with the traditional focus on weapons platforms or individual systems shifting to a systems or enterprise focus. Integration and interoperability became important domains of defense planning, with all manner of "time dominance" performance measures, concepts, and enterprise operational objectives shaping defense planning and military strategy.

Thinking about time-dominance was certainly not new to military strategy. But the advent of information technology in combat decision making coupled with the potential for conflict to escalate into nuclear conflagration altered the importance of the temporal dimension. Time compression issues resonated with a generation of defense planners whose formative years included Pearl Harbor and the 1950s surprise attack discussions. A premium was placed on a commander's ability to surmise the battlefield situation, make decisions rapidly, and adapt battle plans dynamically. Supreme Allied Commander, Europe (SACUER), Alexander Haig, called attention to the temporal dimension in the 1970s, opining that "modernizing conventional forces is our first priority—not because theater or strategic forces are any less important, but because our conventional force deficiencies are the most serious. These deficiencies are exacerbated by trends which, if permitted to continue, portend a diminishing cushion of *warning time*."[31]

Widespread debate occurred about the likelihood of a Soviet surprise attack and, if one occurred, how NATO forces would fare against Soviet armor. A number of analysts did agree that spatial and temporal factors favored the Warsaw Pact—a function of potential combat power generated over a relatively quick period and NATO's lack of operational depth. On this dimension, the advent of the Operational Maneuver Group (OMG) was perceived as a new threat to deterrence stability that rekindled fears of a surprise attack.

The Soviet approach to ground combat involved multiple layers or echelons of combined arms formations successively pushing forward on the battlefield. Breakthroughs were exploited and reinforced using the momentum of second and third echelons. Speed and direction provided velocity; velocity and mass provided combat power to achieve higher-order missions, including opening a salient to flow forces into the enemy's rear. Because second and third echelons were intended to exploit a breakthrough as it emerged they had to remain agile, able to switch direction quickly. This required a formation for movement that enabled agility, known as a column or "march" formation. Lateral dispersion was impractical for follow-on echelons because it would complicate command and control and delay the attack. By the 1980s, the Soviet approach further evolved with the introduction of independent maneuver elements to rapidly penetrate into enemy territory, mixing the lines to lessen the likelihood of a nuclear defense. This meant forward troops would be absorbing Soviet armor with little opportunity to transition from defense to attack.

Chapter 4 discussed the evolution of the Soviet nuclear RMA up to the introduction of the OMG, which became a source of much contention in the West. Western analysts confirmed the OMG's existence and operational characteristics during the summer 1981 "Zapad" (Zapad-81) training exercise in Poland.[32] The OMG was not merely a reinforcement of the second echelon. Significantly more capable, it reflected a resurrection of the Second World War idea of a mobile strike group seeking to create new opportunities for breakthroughs. Such developments in Soviet forces impelled US grand strategy changes concerning conventional readiness. Military analysts noted its destabilizing "shock" and "surprise" power. Lieutenant General Jim King (retired) recalls that the OMG threat "drove" technology initiatives and operational innovations in the European Command, including a number of intelligence advances.[33]

Initial discussions of the OMG's operational mission settled on the extension of the so-called "conventional option" mentioned in Chapter 4: penetration of NATO lines to capture critical targets, disrupt command and control, and prevent the use of nuclear weapons. Presumably, the OMG would encompass an independent maneuver element, perhaps a reinforced division, tasked with breaking through a weak spot in enemy lines to drive deep into enemy territory. Such a drive into the "operational depth" of NATO's territory would disrupt command and control, facilitate the OMG's seizure of critical terrain and river crossing sites and, more importantly, prevent the enemy from launching a nuclear attack on the OMG (as this would mean employing nuclear weapons on friendly territory). Having achieved a penetration, the OMG would be followed by large armor formations. NATO forces would be hit with successive waves of massed armored attacks.

Among the disturbing aspects of the OMG was the implication that the Soviets could wield decisive conventional power more quickly to achieve strategic, theater objectives. US policy makers and Congressional leaders perceived the conventional buildup and Soviet operational innovations as a destabilizing threat that had to be countered. Air assault and helicopter raids conducted in Afghanistan suggested an increase in Soviet confidence to mount conventional strikes.

Additional assessments of battlefield space-time asymmetries reinforced Warsaw Pact advantages. Soviet forces had more depth to mount an attack than NATO did to absorb and prepare defenses, enabling the Red Army to stage echelons in march formation to rapidly flow troops forward. One defense think tank study warned that "a single [U.S.] battalion might find themselves facing 100 to 120 advancing tanks over a 20-minute period" which, given the time required to coordinate defenses, left only two paths for defeating a Soviet attack: either significantly raise the rounds fired per minute "or the number of minutes available" between echelons.[34]

Among the options selected for ameliorating the challenge for Army battalions facing a Soviet armored onslaught was pushing the defense "over 50 to 60" kilometers rather than the 5 to 8 prescribed in early 1970s planning documents. Defending 50–60 kilometers out suggested new spatial dominance requirements for defending forces; temporal dominance requirements followed. "The brigade was responsible for all forces within a distance of twelve hours of the forward line of troops, the division out to twenty-four hours, and the corps to seventy-two hours."[35] Long-range precision strikes and air power were assigned an increasingly important role as planning discussions extended spatial and temporal dominance requirements to 300 kilometers.

Subsequently, military leaders were "told unequivocally of their new responsibilities for the effective management" of all intelligence and intelligence-gathering systems, leading to better integration of intelligence personnel and capabilities into command and control staffs.[36] Only through such integration would the commander be able to "see" his adversary on the battlefield, understand all of the contextual and situational aspects of the battle space, have knowledge of his own and other friendly forces, and engage them with long-range precision fire in time to prevent Soviet armor from breaking through NATO front lines. At the intersection of time and space dominance was the need for precision, especially in terms of being able to deliver weapons to the right coordinates at the right time.

The Air Force slowly increased the emphasis placed on munitions development. A new Armament Division at Eglin Air Force Base, Florida produced a dozen new nonnuclear ground attack weapons in the 1980s. Miniaturization, microchips, and more efficient sources of electro-mechanical power also improved conventional capabilities. Long-range precision strike advances included successive generations of Paveway bombs. The GBU-15 entered service in 1984. An improvement to the electro-optical bomb used in bridge attacks mentioned in Chapter 4, the GBU-15 employed television guidance to glide bombs to targets from distances greater than those feasible with laser designators. GBU-15s also offered pilots and weapons officers the option of locking the weapons onto the intended target prior to their release. A warhead guided by an infrared sensor was fielded in 1987. Another development in precision weapons munitions came from the Navy. Seeking even more standoff range, the Navy added a rocket booster to the Paveway II. Designated the AGM-123 and fielded in 1985, the rocket-assisted precision bomb could be released dozens of miles away and guided to the target via a data link. The AGM-123 delivered a 1,000 pound warhead with astounding accuracy.

Planning strikes to battlefield depths of 150 to 300 kilometers galvanized new thinking about intelligence, command and control, and fire support relationships. Targeting at such ranges required near real time intelligence to be exploited and disseminated to commanders and fire support centers. Targeting now focused on increasing the number of minutes between echelons arriving at the front; fighting by minutes became a more important planning factor than merely thinking about the movement of forces across territory.

General Don Starry's April 1983 testimony before the House of Representatives Committee on Armed Services outlined how advanced conventional weapons systems could be employed to defeat the OMG. The essence of the approach, shown in Figure 5.1, was creating a "window of opportunity" in which enemy forces would be degraded such that the United States could transition to the offensive.[37]

With interdiction, planners believed that a window of time would open in which NATO forces would have an opportunity to halt the attack and stabilize a new front. This reinforced the focus on temporal dimensions of planning, designing new weapons systems, and revising doctrine. The conceptual and doctrinal implications of focusing on specific time windows paved the way for rapid dominance and rapid decisive operations discussions in the 1990s.

Planning had to be more rigorous. Paradoxically, flexibility depended on the rigor inherent in planning. The need for more deliberate intelligence preparation of the battlefield (IPB) forced commanders and battle staffs to conduct detailed studies of terrain and enemy capabilities to prepare mentally, train organizationally as the unit would fight, and to understand the complexities of how resources would have to be orchestrated if commanders were to seize opportunities on the battlefield. Among the new planning support requirements was the need for elevation data and automated location capabilities to rapidly emplace mortars and artillery pieces to support indirect fire support missions.

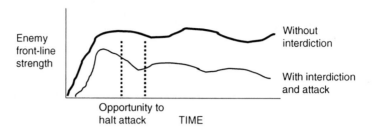

Figure 5.1 Interdiction and attack.

Flexible logistics were also needed, a task that required more dynamic geospatial referencing for combat support units so they could keep pace with maneuver forces. Dynamic air support also became more important, again reinforcing the need for spatial awareness of friendly and enemy units. A premium was placed on simultaneously coordinating the close fight and prosecuting deep strikes.

A related initiative involved the reorganization of Army divisions. In August 1983, Army Chief of Staff General Johan A. Wickham directed TRADOC to examine Army organizational structures in light of emerging global response missions. TRADOC's efforts resulted in an Army of Excellence force structure based on the need for much smaller, more easily transportable divisions with sufficient capabilities to wage limited war. Heavy divisions organized around large armored formations were retained, but light divisions would become the focus of organizational and doctrinal innovation through the end of the Cold War. Armies of Excellence concepts, doctrine, and training requirements, furthermore, were forerunners of the Army's decision in the 2000s to accelerate development of small, flexible, rapidly deployable Brigade Combat Teams.

What had begun in the late 1970s as a quest to integrate weapons systems to offset Soviet conventional superiority quickly led to other types of integration, including integrating nuclear and conventional, a "joint" warfighting framework integrating air, land, sea, and space domains, and integrating the close fight on the front lines with deep attack and interdiction. Much of the original impetus involved spatial dominance, although in the 1980s time dominance became more important. By the 1990s, spatial and temporal dominance requirements evolved into a full-spectrum dominance strategy, with an emphasis on a rapid dominance approach to planning and operations.

Changes in how leaders conceived of their ability to influence the battlefield led to other developments. Commanders focused on seizing and maintaining the initiative. Planning for combat sustained a sense of creativity. Leaders—and soldiers—knew their fate was not completely controlled by higher commands. Leaders, furthermore, were conditioned to expect situational awareness. Commanders focused on finding transition points in battles to shift from the defense to the attack.

Throughout, several types of integration dominated military strategy and defense planning. In the 1970s, the aim was integrating nuclear and conventional warfare and integrating the close and deep battles. In the 1980s, the aim was integrating European strategy with regional contingencies, integrating the Service through joint doctrine, and integrating all forms of national power to exhaust the Soviet empire. Underwriting these and other types of integration was the deliberate integration of information technology into all levels of defense planning and military strategy, from the conceptualization of battlefield operations in terms of time and space dominance through the approach to research and development.

Systems engineering and integration capabilities skills subsequently evolved as key industry skills. Winning the position of a "prime" contractor for new weapons systems now meant being the lead integrator, not the primary developer. Jacques Gansler, Undersecretary of Defense for Acquisition, Technology, and Logistics during the Clinton administration recalls that the 1980s brought increased "prime" awards to companies like IBM, with traditional platform developers like Boeing having to adapt their strategies for winning such contracts by proving their systems integration skills and ability to oversee software development activities.[39]

The culmination of two decades of integration as an overarching theme in defense planning and military thought was the centering of the "network" concept as a grand narrative in military strategy and doctrine.

Military thought and doctrine: AirLand Battle and Follow-On-Forces Attack

During the debate over Active Defense doctrine at the end of the 1970s, then Lieutenant General Don Starry identified two additional problems to those mentioned in Chapter 4. The first concerned the insular drafting process used in 1976. Not only did it exclude the preponderance of Army senior leadership from the coordination process, it did not involve mid-level leaders destined to implement the doctrine in future battles. A second problem concerned criticism that the 1976 manual placed too much emphasis on the defensive. Generals Starry and DePuy had recognized this but had no solution for the problem of defeating successive echelons of Soviet forces. This second problem was revisited during debates over the OMG threat.

As interdiction strategy emerged in the 1980s, FM 100-5 underwent revisions to accommodate the desire for a more offensively-minded doctrine. AirLand Battle, a term coined in 1981, replaced Active Defense as official Army doctrine in August 1982; a revised doctrine came in 1986. The second iteration of AirLand Battle reinforced a new doctrinal emphasis on initiative, maneuver, and joint operations. This, in turn, reflected changes in foreign policy and a new strategic outlook defined by a more assertive American military presence abroad.

In July of 1977, Starry left Europe for Fort Monroe, Virginia, earning his fourth star and assuming command of TRADOC. Soon thereafter he changed the doctrine drafting process to make it more receptive to input from across the Army. He also distanced himself from the process to minimize criticism that the TRADOC commander's own preferences were influencing the final document.

Starry assigned drafting responsibility to Major General William R. Richardson, commandant of the Command and General Staff School and commander of the Combined Arms Center at Fort Leavenworth, Kansas. Richardson guided a handful of writers, among them then Major Leonard "Don" Holder and Lieutenant Colonels Huba Wass de Czege and Richmond B. Henriques. Wass de Czege, who retired as a brigadier general, was the son of a Hungarian writer who fled his homeland in 1956. A career light infantry officer, de Czege is a legend among military thinkers for his unabashed romanticism and penetrating mind.

The architects of AirLand Battle were aware of a weakness in Soviet military doctrine and war planning: a penchant to adhere to precisely scripted movement tables, a rigid command and control system that stifled initiative at local levels, and an educational system that failed to nurture a creative approach. Railways and over-the-road heavy equipment movers brought men and equipment to the eastern end of nine mobility corridors running westward into NATO territory. The echelon construct required adherence to detailed plans—down to regimental levels. Commanders rehearsed battle plans to ensure their arrival at designated points at specific times. Soviet planning reflected a "scientific" approach to combat.

Richardson's team "addressed Starry's concerns about dealing with the second echelon of any Soviet or Soviet-like mechanized attack" and "reinvigorated the basic doctrine, making it more offensive" in nature.[40] Tenets of the new doctrine included initiative, depth, agility, and synchronization.

According to one account, AirLand Battle evolved from defense consultant Joe Braddock's ability to identify "a pattern in the operations, exercises and planning of the Soviet Union and Warsaw Pact forces. This called for particularly close coordination between the Air Force and the Army, to prevent the Warsaw Pact from being successful in the forward areas, primarily through firepower and maneuver, while at the same time being able to coordinate attacks on their rear areas that disrupted their capability to reinforce and influence the action in the forward area."[41]

Although there were already "some seventy battlefield systems and subsystems in various stages of conversion to automation" before AirLand Battle Doctrine was adopted, the new FM 100-5 helped shape thinking about modernization and promoted systems integration. Drawing on the new doctrine, "the Army was eventually able to conceptualize a tactical command and control architecture" for controlling maneuver and coordinating fires, automating some indirect support, and revising approaches to intelligence support (including electronic warfare).[42]

Noteworthy here is the fact that calls for initiative and adaptability increased as the area of a commander's combat influence increased. As mentioned above, AirLand Battle settled at a depth of 150 in front of the forward lines, largely because this is the depth Army and Air Force leaders agreed upon to both protect Air Force interdiction missions and allow the Army to develop its own indirect fire weapons. In addition to extending the area a corps commander would "influence" through deep attack to 150 kilometers (roughly 72 hours in Soviet doctrine); AirLand Battle also set the commanders area of interest—the area for ISR assets to provide information on—some 300 kilometers out (roughly 96 hours). Extending the area of influence and interest established new requirements for tactical reconnaissance, targeting, and strike planning.

The renaissance in American military strategy and operational doctrine that took form in the 1980s was partly a rediscovery of the campaign ideal for theater-level, nonnuclear operational planning. Indeed, the rediscovery of the campaign planning and execution as a high-order form of military science and requirement for command created an environment in which readiness to plan for and engage in conventional warfare was more openly questioned. This led to additional debates about doctrine, logistics, training, leadership, and intelligence support.

One of the most important outcomes was the emergence of a new way of thinking about operations. Luttwak lamented that "Anglo-Saxon military terminology" in the early 1980s addressed "tactics (units, branch, and mixed) and of theater strategy as well as grand strategy, but includes no adequate term for the operational level of warfare" or the art of the campaign—despite recognition that such a level of warfare was a core element of classic military thought.[43] As Ben Lambeth contends in his study of the transformation in American airpower from the 1970s through the 1990s, "US defense leaders not only did not speak in these terms but also did not even think in them."[44]

Arguably, the emergence of operational thinking and an impetus on teaching campaign planning both reflected recognition that the dominant narrative of nuclear thought in US strategic discourse had stifled nonnuclear planning and doctrine and signaled the first steps toward a strategic discourse that accommodated new operational challenges in Europe and other theaters.

Campaign planning had to be taught. The School of Advanced Military Studies (SAMS) opened in 1981 to give Army majors a better understanding of the operational level of war. Wass de Czege was instrumental in creating the school, was its first director, and taught a course on applied military strategy. The SAMS curriculum remains fundamentally about operational art, teaching students about large unit operations and campaigns. It is the world's leading school for educating future officers about what a later discussion will call the "knowledge burden" of warfare in the information age.

The creation of SAMS coincided with an Army-wide turn toward the study of Clausewitz and Jomini. Renewed interest in classical strategic thought was in part engendered by a US Army War College-sponsored study on the strategic lessons from Vietnam. Colonel Harry Summers' study, *On Strategy: The Vietnam War in Context*, drew heavily on Clausewitz's military theory to frame the political–strategic factors of war and their relationship to the conduct of war.[45] He documented a lack of appreciation for military theory and a blurring of the relationship between military and national strategy, in part a result of decades of overemphasizing nuclear strategy.

Professional military journals were inundated with discussions of Clausewitz, the principles of war, and the utility of operational art as a means to rethink planning for and prosecuting campaigns. Debating the principles of warfare and their relevance became a common theme in military science. Within the Army and the Marine Corps, debate rekindled interest in maneuver warfare theory. Meanwhile, Army and Air Force leaders began collaborating, at times reluctantly, on technology and operational concepts to help restore maneuver to ground forces and to integrate air and ground efforts against Soviet follow-on echelons.

Air Force planners viewed AirLand Battle suspiciously. For many pilots, the emphasis on deep strike interdiction against Soviet armored forces threatened to turn their aircraft into flying artillery platforms to support ground troops, detracting from their independent strike mission and exposing them to enemy surface to air missiles. Moreover, implementation of AirLand Battle would seemingly subordinate airpower to the main ground effort, with Air Force capabilities treated as direct support to Army interdiction capabilities. Some emerging Army capabilities, including long-range missiles with terminally guided submunitions and land-based cruise missiles, challenged traditional divisions of battlefield responsibility for long-range precision strike. Army leadership went to great lengths to assuage Air Force critics that AirLand Battle treated ground and air power as co-equal elements of the overall effort. Growing emphasis on Army deep strike missiles and Army attack helicopters, however, led many in the Air Force to remain skeptical.

In the midst of an ensuing "free-swinging doctrinal debate," Allard writes that Army Chief of Staff General John A Wickhman, Jr and Air Force Chief of Staff General Charles A. Gabriel found themselves united by long-standing personal friendship and remarkably similar viewpoints on the need for closer cooperation between the services they led."[46] High-level support for additional cooperation came in the spring of 1981, when Army and Air Force leaders agreed to a Joint Suppression of Enemy Air Defense (J-SEAD) project that resulted in subsequent, historic Army–Air Force agreements to share long-range interdiction missions. Arrangements were made for Air Force control of deep strikes using Army missiles and Army management of the prioritization of air strikes on second-echelon Soviet armor.

In mid-1983, Wickhman and Gabriel "quietly put their staffs to work on a cooperative project to rationalize the planning and development of joint combat forces centered around the AirLand Battle model."[47] Jointness—or rather the lack of it—became the overarching defense readiness issue after the invasion of Grenada. Communication and other interoperability issues received national media attention, suggesting that Reagan defense initiatives might be building a force unable to fight effectively. At times a rancorous process, and despite failure to resolve important issues, this period was a turning point in the evolution of joint warfighting. Operational and political exigencies eventually brushed aside some of the organizational impediments to collaboration and cooperation by highlighting the downsides of *not* developing a joint warfighting approach. Air Force resistance to an Army deep strike role receded as the Air Force assumed new missions, including a lead role in space-based support to military operations. GPS and communications were key concerns because they were critical for the coordination of effective deep strike missions.

Although not perfect, Army–Air Force cooperation also improved during this period as Services revisited conventional war planning requirements. Both sought to dominate adversaries in all types of weather and at night, yielding a decisive American advantage by the end of the decade. They also collaborated on plans to simultaneously defend and attack throughout the depth of the battlefield. The close-in fight would be coordinated with deep attack, creating opportunities to blunt a Soviet attack through interdiction and maneuver.

Army and Air Force leaders worked on policies and operational responsibilities for AirLand Battle implementation from November 1983 through March 1984. Discussions resulted in the Army and Air Force Chiefs of Staff agreeing to the so-called "31 Initiatives" on roles and missions. The 31 Initiatives built on an earlier 1981 agreement on offensive air support, which transferred limited planning authority over close air support, the tasking of strike aircraft for battlefield interdiction missions, and collection planning for airborne reconnaissance to ground force commanders. An important aspect of the agreement was cooperation on J-STARS development.

Meanwhile, NATO planners and European defense analysts turned their attention back to evolving interdiction requirements. Advanced target acquisition and long-range strike capabilities were increasingly perceived as an alternative to nuclear weapons.[48] A 1983 European Security Study group's report entitled *Strengthening Conventional Deterrence in Europe*, known as the ESECS report, recommended a deep attack strategy and operational capabilities to attack 300 kilometers beyond the forward line of NATO forces. The ESECS report also called attention to widespread agreement among technology experts that the required conventional capabilities could be fielded in five years. A year later, a British Atlantic Committee report entitled *Diminishing the Nuclear Threat* reinforced arguments for advanced conventional capabilities. As envisioned in the Offset Strategy, technology would enable both a more effective defense and the ability to transition to the offensive to attack targets throughout the Soviet's operational depth.

NATO doctrine also changed in the 1980s. By the end of the decade, NATO adopted a strategy of Follow-On-Forces Attack (FOFA), the name given to a concept made public by Supreme Allied Commander, Europe (SACEUR) Bernard Rogers. Rogers began thinking of the concept in 1979, around the same time that Army leaders decided to replace the widely criticized 1976 Active Defense doctrine. FOFA, which some referred to as the Rogers Plan, differed from AirLand Battle in some operational details but complemented it as an overall approach. Both built on a multifaceted view of integration and sought to achieve battlefield extension by carrying the defense of NATO forward into Warsaw Pact territory.

Rogers had assumed command of NATO forces at a time when the only viable counter to Warsaw pact forces was nuclear weapons. Based on an understanding of emerging technology, including the Assault Breaker demonstration program, he conceived of an ability to detect and respond to a Soviet attack with thousands of deep-strike missiles and rockets. The impetus for FOFA was "reducing to a manageable ratio with conventional weapons the number of enemy forces arriving at" NATO's forward defense lines, its "General Defense Position."[49] Similar to AirLand Battle, FOFA aimed to simultaneously disrupt and attack the first and second echelons of Soviet forces as an alternative to using nuclear weapons.

As discussed earlier, a key tenet of both AirLand Battle and FOFA was applying interdiction capabilities to elongate the periods of time between arrivals of successive echelons at NATO's forward defensive positions. If these periods could be increased, NATO's forward forces could regroup, consolidate defenses, and in an ideal situation push forward. Merely creating time windows or pauses, however, was not enough. Information from across the front had to be gathered, analyzed, and exploited. In knowledge management terms, information (the "know what") had to be exploited to create knowledge about enemy vulnerabilities and opportunities to turn the tide of battle (the "know how").

As did the Offset Strategy, AirLand Battle and FOFA sought to leverage technology and doctrine to increase the power of existing weapon systems and restore deterrence stability by providing a credible defense and retaliation alternative to the first use of tactical nuclear weapons. In addition to target acquisition, command and control, and guidance systems, new technology promised to improve overall warning of attack and to generally increase the accuracies of long-range strikes. They also both required better coordination and collaboration between air and ground forces. Army–Air Force cooperation reportedly helped reduce duplication among the two services and enabled some "$1 billion in associated savings."[50]

Some criticized AirLand Battle and FOFA for relying on immature and costly technologies. Others contended that relying on deep strike instead of additional frontline defenses would leave forward forces vulnerable to OMG and other penetration units, opening opportunities for subsequent echelons. Analysts concerned with Soviet reactions and potential escalation questioned how Soviet theater nuclear commanders would discriminate FOFA conventional missiles from nuclear ones. FOFA was also criticized as further evidence that the United States wanted to lessen reliance on nuclear weapons, a potential weakening of the credibility of NATO's nuclear deterrent. Still, the operational ideas and a shift toward reliance on information systems marked a turn in thinking about warfare.

As AirLand Battle and FOFA shaped thinking about operations, the Air Force adapted the role of the bomber in theater and tactical situations. After an internal debate over the imbalance between strategic and tactical capabilities, the Strategic Air Command's B-52 squadrons increased training for conventional combat missions. It was also the era of the so-called "fighter mafia," a cohort of fighter pilots and conventional force planners who argued for rethinking the core mission area for Air Force. No longer would air–ground attack, close air support, and air superiority missions be subordinated to strategic bombing.

This required more timely information on the location of enemy forces. For an Air Force coming to grips with the complexity and rapidity of operations, this called for richer information context over greater distances in less time. Updated enemy situation data and geo-positioning for weaponeering were two important developments during the 1980s that built on digital, softcopy information sharing and command and control advances. Specific information technologies contributing to increased military effectiveness for both air and ground forces included: digital imaging from spy satellites, analytic stereo-photogrammetry for targeting and other precision location missions, global positioning system applications, precision terminal guidance through scene-matching correlation, high-bandwidth secure digital battlefield communications, automated information fusion and analysis, hardened (jam resistant) command and control, stealth planes, and cruise missiles.

Organizational developments included more realistic training using digital terrain elevation data, closer integration of intelligence production and analysis with operations staffs, refined decision processes using near-real time dissemination of intelligence and cartography, increased joint warfighting, heightened Congressional involvement in acquisition planning, and processes to consider the costs of information needed to make new weapons work as part of procurement decisions.[51]

One of the most important outcomes of doctrinal innovations in the 1980s related to planning and joint operations was recognition that the joint command structure needed reform. Before retiring in 1982, Chairman of the Joint Chiefs of Staff General David Jones criticized the existing joint service structure in testimony before the House Armed Services Committee. Among the deficiencies Jones noted were the dilution of military advice because each of the Service Chiefs had an equal voice in decisions and the weakening of the Combatant Commander's authorities to plan for and conduct operations in their respective regions or functional mission areas. A series of studies in the early and mid-1980s added recommendations for joint reform and fueled debate. The Senate actually passed a bill in 1986 calling for the replacement of the Joint Chiefs with a panel of senior advisors; the House amended a defense authorization bill with less extreme reforms. The Goldwater–Nichols Reorganization Act of 1986, the compromise legislation that emerged from the Senate and House proposals, was signed by President Reagan on October 1, 1986.

Goldwater–Nicholas proved itself among the most important military reforms of the Cold War. In addition to designating the Chairman of the Joint Chiefs the principal military advisor to the President and Secretary of Defense, it gave the Chairman additional authorities for budget coordination, strategic planning, force structure recommendations, and coordinating logistics. It also clarified the operational chain of command and increased the power of the Combatant Commands. The military services were clearly assigned the mission of training, equipping, and organizing military forces and the combatant commands were given operational command and control over the forces assigned to them. Additional reforms strengthened joint education and training, joint career progression, and responsibilities for representing the combatant commands in programmatic decisions.

New security challenges and military missions at the end of the Cold War

Two security issues became more important in the 1980s: preserving Persian Gulf stability and fighting international terrorism. Together, these issues created new missions for the armed forces that reinforced developments in American military thought, military doctrine, and defense planning. The evolution of thinking about rapid deployments and rapid dominance, for example, were influenced by the need to plan for Gulf contingencies and to deploy teams clandestinely around the world. The need to plan for contingencies outside of the European theater of operations raised new questions about logistics, mobility requirements, basing rights, and training. Additional requirements addressed intelligence, surveillance, and reconnaissance capabilities and a more sophisticated global command and control infrastructure.

The offensive spirit permeating new operational concepts and training programs devised for deep battle operations under compressed timelines in Europe was transferred into other contingency planning. Underscoring the new offensive doctrine was a sense of urgency to win the first battle of any future war and to never relinquish the initiative, a mindset that imbued new planning tasks with a sense of confidence. Such factors certainly shaped the initial approach to Persian Gulf planning, which posed different spatial and temporal challenges due to the lack of American military forces in the region, uncertain allies and security partnerships, and the fact that Soviet forces were much better positioned to engage militarily in the region. Because the region was critical to Reagan's strategy of weakening the Soviet empire, however, confidence morphed into innovation. Persian Gulf planning accelerated the incorporation of rapid dominance and rapid deployment concepts into American military thought and defense strategy.

Recall Chapter 4's discussion of Soviet expansion into the Persian Gulf during the 1970s and the December 1979 invasion of Afghanistan. In the 1980s, some feared Soviet subversion in Pakistan or an attack into Iran and toward Gulf oil fields from the foothold Soviet forces had secured in Afghanistan. Carter increased aid to Pakistan, announced the Carter Doctrine on January 23, 1980, and increased military spending to expand the size and quality of the American conventional warfighting force.

Reagan altered the pace and scope of the military buildup and he intensified the level of American military activities abroad. US foreign policy scholar Cecil Crabb argues that Reagan Secretary of State Alexander Haig's 1981 visit to the Persian Gulf marked an evolution in the strategic intent underpinning the Carter Doctrine. It indicated an acceptance of the "Carter Doctrine as an axiom of American diplomacy in the Middle East" and signaled that "Republican policy-makers accorded the preservation of Middle Eastern security from Soviet hegemony an even higher priority."[52]

Oil was of course a strategic concern. "If the industrial democracies are deprived of access to those resources," Harold Brown argued in February 1980, "there would almost certainly be a worldwide economic collapse of the kind that hasn't been seen for almost 50 years, probably worse."[53] The 1970s OPEC oil embargo and a decade of crude oil price fluctuations helped usher in a new era in security affairs in which the intricacies and interdependencies of a globalized world economy were elevated alongside the global nuclear balance of power in security policy decisions. Moreover, economic security required a different type of stability, one that could not be addressed credibly with a nuclear deterrent.

During the last years of the Carter administration, CIA analysis predicted that Soviet oil production would experience a sharp downturn in the 1980s. The economic consequences of declining oil production would create enormous political pressure to seek alternate sources through military force, political coercion, or diplomacy—all to the detriment of US interests. Reagan administration officials feared that Soviet oil shortages would lead Moscow to seize oil fields or, more likely, coerce oil-producing states into selling oil for rubles (not convertible in the world market) or exchange oil for Soviet military equipment (leading to increased presence of Soviet military advisers). Global economic chaos would result. Analysts concluded that the Soviets could place some "four airborne divisions, four surface-to-air missile units, and one motorized rifle regiment" within five days and an addition motorized rifle regiment "every 27-to-48 hours thereafter."[54]

Persian Gulf contingency planning complicated the national security planning process at a time when the American defense establishment was already stressed. Despite confidence in American forces, regional contingency plans suggested new readiness shortfalls. Exercises to refine contingency plans exposed inadequacies in the planned force structure, including logistics, sealift, and other support capabilities. As American security commitments in the Persian Gulf and other regions increased, so too did American military deployments and the frequency of training exercises. All of this suggested a level of assertiveness not present in the immediate post-Vietnam years.

New planning challenges emerged for the US military as planners adapted principles and practices conceived for rapid reinforcement of NATO to rapid deployment into areas where the United States did not have permanent operating bases. Decades of focus on nuclear contingencies and planning for the reinforcement of NATO with heavy armor units created organizational and conceptual inertia that impeded new thinking about missions to protect oil fields in the Gulf. Projecting military power into the Persian Gulf region was no easy task. The region was more than 6,000 air nautical miles and 8,000 sea nautical miles from the United States and the United States had few military bases from which to stage operations. Insufficient combat forces were committed to provide the resources needed to project power into the region. Mobility forces (sealift and airlift) were particularly lacking.

As in Europe, the United States faced time-space asymmetries favoring Soviet forces. Logistics remained a concern for both the European theater and the Persian Gulf. During the November 1981 "Bright Star" exercise it took some four days to transport four hundred men across the Atlantic. Military equipment had to be shipped in a West German freighter because of inadequate military sealift, one that required weeks of advanced notice to ensure availability. That same year defense planners simulated Soviet invasion of Iran to assess US response capabilities. The results confirmed logistics and mobility shortcomings.

Staging the required equipment would take six months. Analysis of rates of munitions usage during the 1973 Yom Kippur War suggested that planned ammunition stocks and pre-positioned reserves might be exhausted before re-supplies arrived. Analysis of the volume of potential fire missions that could occur on a flat, desert battlefield complicated logistics planning because the original planning for a rapid deployment force depended on units that were normally assigned a European mobilization mission.

Bright Star did lead to some positive lessons. US power projection into the region was demonstrated by B-52 bombers flying non-stop from North Dakota bases, a training mission the same units would operationalize in the 1991 Gulf War, in the removal of the Taliban from power in Afghanistan, and the 2003 invasion of Iraq. Another result of Bright Star was an expansion of the POMCUS (pre-positioned overseas material configured to unit sets) program to enable pre-positioning of three additional divisions by 1983 and three more by the end of the decade. Military exercises also addressed the defense of Oman and the Sudan.

Planning for regional contingencies received renewed impetus when Reagan transformed the Rapid Deployment Joint Task Force (RDJTF) into the US Central Command on January 1, 1983, the regional combatant command that would lead the 1991 Gulf War and the 2003 invasion of Iraq. Its first commander was then Lieutenant General (LTG) Bob Kingston (part of the trio who helped develop the concept for Delta Force). Central Command's objective was "to disrupt at their outset the attacks of Soviet or Soviet-client forces, and control the battlefield environs for the time required to deploy US reinforcements and re-supply from distant points" into the theater.[55]

To project a credible force, planners realized that the United States would have to keep mobilization equipment in the region, prepare airstrips and support facilities for additional deployments, and rotate other forces—carrier battle groups and Marine detachments—into the region on a routine basis. Reinforcements and deploying forces would have to be trained to fight "light," a requirement that spurred additional force structure planning to provide an array of lighter armored vehicles than those being fielded for European contingencies. Additional requirements surfaced for indirect fire capabilities that could be rapidly deployed and readied for operations in austere operating environments.

Armored forces dominated thinking about future warfare until the 1980s. Light forces were not high on the post-Vietnam priority list. Even within the Marine Corps, the tendency was toward heavier forces able to engage Warsaw Pact armor. Partly this reflected the principle planning challenge: defeating Soviet forces in Central Europe. It also reflected beliefs that light forces were too easily embroiled in regional conflicts. Security challenges in the 1980s, however, contributed to newfound interest in light forces. Army planners recognized the need to provide organizational and technological responses to increased involvement in low intensity conflict, peacekeeping, and nation building. The defense establishment also recognized the need for forces able to counter terrorism and to engage in third world skirmishes.

No history of the thirty-year transformation in US military thought and defense planning can overlook the story of how counterterrorism and special operations forces evolved from marginalized units lacking political support to exemplars of the new style of American warfighting. Terrorism was certainly not a completely "new" security concern in the 1980s. What was different? Terrorism in the 1960s and for most of the 1970s was fairly localized in its manifestation of indiscriminate violence, only becoming "transnational" terrorism at the end of the decade. And while terrorist tactics were not necessarily new, the rise of state-sponsored terrorism was. So too was the increased number of indiscriminate attacks to cause mass casualties. Terrorism also became more effective and lethal when states increased their training, equipping, and financial sponsorship. During the 1980s, moreover, terrorist attacks in Europe increased along with the level of violence against civilians.

Political violence and regional stability concerns were a growing source of instability for US security planners in the 1980s. A wave of terrorism in Turkey, for example, led to a military takeover in September 1980. More ominous in retrospect because of its long-term implications for the germination of a radical Islamic ideology, the Muslim Brotherhood joined forces with secular groups in Syria and Egypt, leading to new models of anti-Western opposition. Islamic militants opposing the Soviet invasion of Afghanistan were among the users of new communications technology, employing it to attract waves of recruits who received their training in Pakistani religious schools and training camps. Thousands of fighters flowed through these schools and camps where the struggle against Soviet invaders provided unity of purpose. Global communications technology provided terrorists with access to new media outlets, gaining a larger audience for ideologically motivated violence.

Marines were deployed to Beirut in September 1982 as peacekeepers with a multinational force to stabilize the central government and end nearly a decade of violent civil war. The October 23, 1983 bombing of the Marine barracks in Beirut, Lebanon was the tipping point in US planning to combat terrorism. Only a month earlier terrorists had attacked the American embassy. The killing of 241 US Marines in Beirut stirred US action. After the American embassy in Kuwait was attacked in December 1983, planning began for retaliation. Meanwhile, a wave of terrorist attacks around the world brought a new sense of American vulnerability, one that added to the lingering sense of helplessness felt during the Iranian hostage crises.

Although Reagan did not retaliate immediately, the Beirut bombing shook the national security establishment. NSDD-138 was signed in April 1984, endorsing "preemptive raids and retaliatory strikes" against sponsors of terrorism and terrorists themselves; it also "ordered twenty-six federal departments and agencies to develop plans for combating terrorism."[56]

In 1984 there was a significant increase in the number of Arabs arriving in Pakistan and Afghanistan to join the Afghan Mujahideen in their jihad or holy war against Soviet occupation. The Taliban, drawing its name from the Arabic word for student, emerged as a unified group with a common core of intellectual thought based on shared experience in Pakistani refugee camps and madrassas (religious schools). Two years later the Mujahideen defended the mountain village of Jadji from a Soviet attack. Osama Bin Laden participated in the operation, which lasted seven weeks and ended in Soviet withdrawal. Following the battle of Jadji, "Saudi Arabian Airlines gave 75 percent discounts on flights to Peshawar to men going to join the Mujahideen. At times, the Pakistani embassy in Riyadh was delivering up to 200 visas to the young recruits."[57]

Libya, the leading supporter of terrorism in the early 1980s, became the focus of attention. A 1985 Special National Intelligence Estimate concluded that Libyan strongman Muammer al-Qaddafi was supporting terrorism or insurgency in some twenty-five nations across the globe and backed opposition groups in a dozen others.[58] Libya ran dozens of training camps. When Libya was found responsible for the April 1986 bombing of a West Berlin discotheque frequented by US soldiers, Reagan approved military retaliation. In a strike code-named "El Dorado Canyon," Air Force and Navy aircraft attacked Libyan terrorist training camps and other targets in Tripoli and Benghazi. The 1986 bombing of Libya was the first real use of precision bombing since Vietnam and an early case study in the use of American airpower to defeat sophisticated air defense systems. At the time, only Moscow and Baghdad had more advanced air defense capabilities.

As the Cold War was ending, American military forces were being deployed for regional contingency operations, including an increased role in Persian Gulf stability. In 1986, Kuwait asked the United States to place some of its oil tankers under an American Flag in the hopes that doing so would deter Iran from attacking them. The United States launched operational Earnest Will, which resulted in American naval forces destroying Iranian oil platforms being used to coordinate attacks on shipping and eventually sank the bulk of the Iranian navy.

In December 1989, the United States launched a regime change operation to remove Panamanian dictator Manuel Noriega from power. Named Operation Just Cause, the deployment of American forces to Panama marked a turning point in the history of US military operations. There were very few large-scale American military operations during the Cold War. From 1945 through 1985, there were arguably only seven major military operations: Korean, the Berlin Airlift; two deployments to Beirut, Vietnam, the Dominican Republic, Grenada, and naval operations to protect oil shipments in the Persian Gulf region. American military forces were used much more frequently from 1985 through 2005: Libya, Panama, Somalia, Rwanda, Haiti, Bosnia, Kosovo, Afghanistan, Kurdistan, and the Persian Gulf.

Throughout each of these contingency operations, American military forces honed an offensive, rapid dominance approach to warfighting relying on superior intelligence and information capabilities, precision strikes, strategic mobility, and other nonnuclear capabilities initially developed to offset the Soviet military threat.

Chapter conclusion

By the early 1980s, many of the innovative weapons systems conceived as part of the Offset Strategy had reached prototyping and demonstration stages, leading to greater understanding of their potential to alter the strategic military balance and to lessen reliance on tactical nuclear weapons. Arguments for conventional force modernization fell on more receptive ears within the American defense community as the political context shifted. Planners continued to evaluate the credibility of NATO's nuclear deterrent as political opposition to nuclear weapons increased in Europe and the United States.

Reflecting on nearly forty years of defense planning that considered "*strategic warfare*" as "synonymous with *nuclear warfare*," in 1983 defense analyst Carl Builder noted "a perceptible shifting of favor away from nuclear weapons, toward advanced conventional weaponry."[59] Builder conjectured that new conventional warfighting capabilities, "nonnuclear weaponry," would soon be capable of performing "major military roles" on the battlefield that were previously "served by strategic nuclear forces."[60]

Soviet military analysts recognized revolutionary changes in warfare before many of their western counterparts. Soviet analysts, in fact, argued in the late 1970s and early 1980s that American conventional forces were developing precision strike capabilities able to achieve battlefield effects similar to tactical nuclear weapons. Soviet Marshal Nikolai V. Orgarkov, for example, claimed in 1984 that US conventional forces would soon "make it possible to sharply increase (by at least an order of magnitude) the destructive potential of conventional weapons, bringing them closer, so to speak, to weapons of mass destruction in terms of effectiveness."[61] This meant that conventional forces could impede the movement of armor formations and disrupt command and control without threatening nuclear escalation. This had been, of course, part of a larger national strategy that aimed to exhaust the Soviet Union and end the Cold War. "Warsaw Pact defense ministers," Christian Nunlist later learned from Soviet archives, did indeed view "developments in conventional armaments in the early 1980s as more ominous than the strategic change" wrought from nuclear weapons developments.[62]

Soviet reactions to American military modernization included, paradoxically, the first discussions of an American revolution in military affairs (RMA). Soviet perceptions of American conventional warfare innovations had an amplifying effect on US defense modernization decisions in the 1980s. Director of the Office of Net Assessment Andrew Marshall relates that, upon learning of Soviet concerns about American "reconnaissance-strike" initiatives, US defense planners "concluded that it would be useful to intensify those concerns by further investment" in precision strike.[63] As we learn in Chapter 6, decades of Soviet writings on the nuclear revolution in military affairs were later adapted into the predominant theme in the 1990s' defense strategy discussions.

Former Under Secretary of Defense for Strategy and Threat Reduction Ted Warner, for example, notes that US views of the RMA were "heavily based on Russian or Soviet conceptions."[64] Admiral Bill Owens (retired) is among those attributing early recognition of the potential for information technology to underwrite an RMA to Soviet observers. Soviet "technocrats," he posits, first recognized "that computers, space surveillance, and long-range missiles were merging into a new level of military technology, significant enough to shift the balance of power in Europe in favor of the United States and NATO."[65]

New security requirements to deter conflict outside of Europe and defend non-NATO security partners renewed criticism of the 1976 Army field manual, particularly its emphasis on defense. What was different? Two factors stand out. First, the new doctrine pushed ground–air cooperation in new ways. Second, the doctrine began pushing yet another turn toward "speed" within military history. Now, speed involved leveraging information technology to expedite decision making over distributed forces, sustain required levels of lethality, and offset the loss of protection incurred by reducing mass. Further doctrinal changes ensued based on both factors, with time–space compression and expanded, collaborative control over distributed forces reinforcing ground–air cooperation.

Winning the opening battle of any future conflict remained a prominent theme in American military strategy. Yet as the Cold War ended it was unclear what the nature of the next conflict might be. Charles E. Heller and William A. Stofft concluded their classic *America's First Battles, 1776–1965* with a simple observation: "America's ability to predict the nature of the next war (not to mention its causes, location, time, adversary or adversaries, and allies) has been uniformly dismal."[66] Although the United States had planned for conflict in the 1990–1991 Gulf War was indeed a surprise. And despite winning the opening battle, which was the only battle in the first Gulf War, the United States did not prepare for the war that would come a decade later. The first Gulf War was resounding American victory that taught the wrong lessons for the opening battles in the war against violent extremism.

Nations rarely learn from victories as much as they do from failures. This is an important point for American military planners struggling to find a way to victory in Iraq. What American defence planners did not do in the 1990s was refocus modernization to address what many argued would be an era marking the end of large-scale armoured warfare. Too few American defence planners learned the lessons that future adversaries took from US military performance in the first Gulf War. That is, that the only way to beat the American military was by either using weapons of mass destruction or by waging asymmetric war. In the case of the latter, the United States had run too far from the Vietnam experience, failing to consider the need to master counterinsurgency and urban warfare.

Non-traditional or irregular warfare has been a long-standing weakness of the American armed forces—one not addressed during the slow post-Vietnam revitalization. At a 1982 international conference on terrorism and "low-level conflict," for example, then Army Chief of Staff General Edward C. Meyer called attention to a similar tension between planning for low-level conflict (including counterinsurgency) and planning for the dominant planning scenarios, nuclear and large-scale conventional conflict. Then, as now, defence planners tended to make the "greatest intellectual and physical investment in preparations for the levels of conflict we have least often faced."[67]

Similar sentiment can be found among military thinkers in the 1960s as the United States increased its military presence in South East Asia and in the 1980s as forces deployed to Central and South America. Paradoxically, Eliot Cohen observed, while the "most probable loci for American small wars" in the mid-1980s were Central America and the Persian Gulf, the US regional military commands responsible for these areas had "the fewest forces assigned to their peacetime control for training and operations."[68]

A related theme concerns military doctrine. In the early 1960s, the US adapted force structure and military doctrine to reflect President John F. Kennedy's national strategy of flexible response, ostensibly to diversify available response options to Soviet aggression outside of Europe. The prevailing massive retaliation doctrine was losing credibility as a deterrent. Unconventional warfare capabilities, including Special Forces, were developed. In the face of communist-backed guerrilla warfare, Army Chief of Staff General George Decker opined that "any good soldier can handle guerrillas" and the intellectual father of flexible response, General Maxwell Taylor stated "any well-trained organization can shift the tempo to that which might be required in this kind of situation."[69] Counterinsurgencies were generally deemed a lesser-included scenario, meaning that forces trained for combined arms combat in Europe could be shifted as needed to third world conflicts with minimal retraining. A similar syndrome seemed to inflict defense planners in the 1990s.

Cohen states the underlying problem: "An intellectual comprehension of the demands of small wars does not necessarily translate into implementation of the policies required to wage it successfully."[70] Operationally, according to a former commander of US forces in Central and South America, when it comes to mustering resources for small wars, as a nation the United States does not "understand it and as a government we are not prepared to deal with it."[71] Three decades later, the first post-Cold War revision of the US Army field manual on low-level conflicts—renamed *Operations Other Than War*, optimistically concluded "U.S. neutrality in insurgencies 'will be the norm.' "[72]

In Operation Iraqi Freedom, optimistic projections seemingly discounted US forces being pulled into a counterinsurgency struggle. Arguably, Deputy Under Secretary of Defense Paul Wolfowitz himself understated how far off these assumptions were three months after President George W. Bush's May 1, 2004 declaration that major hostilities had ended: "Some important assumptions turned out to underestimate the problem."[73] The realities of counterinsurgency warfare continue to rub against mainstream US military thought and defense planning. In small wars, Cohen opines, "the attractive notion of a violent but brief conflict is chimerical."[74]

If at the low end of the conflict spectrum the United States has a cyclical tendency to forget the exigencies of counterinsurgency warfare, one wonders how shifts at the higher end will play out. Certainly, strategists recognize that overwhelming US conventional superiority motivates potential adversaries to develop weapons of mass destruction. Just as the United States developed first an atomic and then nuclear deterrent to Soviet conventional superiority, others will certainly seek to offset current US advantages. Andrew Krepinevich and Steven Kosiak, for example, warn of a "danger that the development of an effective nonnuclear strategic strike capability by the United States—because it would appear much more usable than a nuclear strike capability—would increase the incentives for potential adversaries to acquire at least a small nuclear arsenal."[75]

6 From RMAs to transformation

Rediscovering the innovation imperative

When students of American defense policy asked at the end of the Cold War, "Where should U.S. armed forces go from here?" the response, generally, was building on "revolutionary" capabilities attributed to US forces during the 1990–1991 Gulf War. During the first Gulf War, American-led coalition forces defeated an Iraqi military of several hundred thousand in less than six weeks with only 240 casualties. Stephen Biddle observes that "this loss rate of fewer than one fatality per 3,000 soldiers was less than one tenth of the Israelis' loss rate in either the 1967 Six-Day War or the Bekaa Valley campaign in 1982, less than one twentieth of the Germans' in the blitzkriegs against Poland or France in 1939–40, and about one-thousandth of the US Marines' in the invasion of Tarawa in 1943."[1]

Defense analyst Ken Allard posits that American military success in the first Gulf War "was the culmination of more than two decades of post-Vietnam renewal" and the product of "an investment strategy that had consciously sought to offset enemy strengths with technological expertise."[2] Former Secretary of Defense William Perry views the 1991 victory as the first and only "test" of the systems built in the 1970s specifically to achieve the offset strategy. From his perspective, the RMA was merely the offset strategy renamed.[3] Noted strategist Edward Luttwak does not make a direct comparison with the offset strategy but does comment on its Deep Attack elements. Luttwak found the reconnaissance-strike capabilities at the core of the offset strategy to be "concrete expressions" of the RMA visions capturing the attention of "military bureaucrats on both sides of the Atlantic at the turn of the millennium."[4]

Numerous developments contributed to the American-led victory in the desert and to arguments about fundamental shifts in warfare. Many of these developments deserve the status of revolutionary advances when considered within their respective domains or areas of operation. GPS revolutionized not only navigation but all activities requiring precision location. Near real-time sensors and intelligence capabilities did provide unprecedented situational awareness to commanders. Aerial refueling fundamentally altered the operational capabilities of strategic airlift, theater mobility, and other dimensions of air power. Widespread use of night vision and night movement shifted the American military's preferred timing of large military movements and multi-unit attacks from daylight to night and from sequential movements to simultaneous ones. A systems approach to warfighting changed expectations for the pace and scope of technology 'integration. Most importantly, a revolution occurred in how armies prepared and trained individuals and organizations for the battlefield.

Noteworthy was the fact that the 1991 Gulf War was planned and led by Vietnam veterans; was won by an all-volunteer force that was professionally trained using revolutionary training regimes; was led at the tactical level by a new generation of American military officers that embraced an offensive doctrine; and, was waged with information systems providing exquisite battlefield situational awareness and decision support.

Among the most frequently discussed battles of the first Gulf War began on January 29, 1991. A column of Iraqi armored forces moved from southeastern Kuwait and occupied Al Khafji, an abandoned coastal town in Saudi Arabia. A second Iraqi armored force was detected the following day, apparently preparing to reinforce Al Khafji and use it as a stalwart to engage Allied forces in ground battles along the Saudi coast. An E-8 Joint Surveillance Target Attack Radar System (JSTARS) aircraft was diverted from its scud-hunting mission in western Iraq to support an air attack on the Iraqi armor. The battle of Khafji lasted roughly a day and a half, resulting in the destruction of some 600 armored vehicles (tanks, armored personnel carriers, and mobile artillery).

As Ben Lambeth concludes, "the combination of real-time surveillance and precision attack capability that was exercised to such telling effect by air power against Iraqi ground forces at Al Khafji and afterward heralded a new relationship between air- and surface-delivered firepower in joint warfare."[5] "The real hero," he posits, was the ability to integrate ground and air operations using the E-8 JSTARS aircraft.[6] Recall that JSTARS traced its roots to 1970s development projects, including the Air Force's Pave Mover radar and the Army's Stand Off Target Acquisition System (SOTAS), which were brought together into a single program office in 1982.

Brought into the Kuwait theater of operations only two days before the air war began, the E-8 was still in its development stage. E-8s were reportedly deployed based on the recommendation of Army Lieutenant General Fred Franks, commander of the VII Corps, who first experienced their operational capabilities during an exercise in Germany. During exercise Operation Deep Strike, JSTARS detected and targeted a Canadian unit playing the role of a Soviet armored column, "achieving 51 'tank kills' as a direct result" and impressing Franks so much that "he later raved about the capability" to General Norman Schwarzkopf.[7] Lambeth's affirmation of the E-8's contribution was widely shared by others assessing American air power in Desert Storm: "JSTARS redefined the meaning of using real-time battlespace awareness to make the most of a casebook target-rich environment."[8]

What emerged from all of this? Not merely a single sensing system: an entire new way of sensing, acting, and achieving on the battlefield by leveraging the power of information technology. By the early 2000s, defense analysts heralded a new American way of war. As longtime *Washington Post* military affairs reporter Vernon Loeb aptly stated, "It took years, and increasingly impressive proof on the battlefield, before these inspirations were recognized for what they were—a new way of fighting that would change the calculations of war and peace in unprecedented, and still uncertain ways."[9]

Since the first Gulf War, US strategic planning and military thought have been inundated with arguments that American armed forces exhibit a significant, discontinuous increase in military effectiveness, "the process by which armed forces convert resources into fighting power."[10] During Operation Iraqi Freedom, the second Gulf War that began with the 2003 invasion of Iraq, and continuing into a prolonged counterinsurgency and national building effort, discussion of revolutionary American military capabilities have been more circumspect than they were a decade ago.

This chapter traces the ascent and evolution of American military strategy from the end of the Cold War through the early 2000s. During this time, defense strategy evolved from RMA discussions to a focus on defense transformation. Along the way, arguably, defense analysts missed an opportunity to examine the military innovations that led to the "revolutionary" capabilities demonstrated by American military forces in the early 1990s. After outlining post-Cold War defense modernization discourse, this chapter argues for additional military innovation studies to inform defense transformation.

Revisiting the American RMA

Andrew Marshall, Director of the Department of Defense's Office of Net Assessment, sponsored the first official study of the changes in warfare being attributed to American performance in the 1990–1991 Gulf War. Two of the research questions shaping the 1991 study remain central to defense transformation planning in the 2000s: How does one identify appropriate innovations when designing a force to meet future threats? How does one foster innovation inside the defense department and military services?[11]

The task fell to Andrew F. Krepinevich, who assessed the overall discourse of change, explored arguments Soviet military analysts were making about American conventional warfighting advances, and detailed how new conventional strike capabilities were increasing the overall effectiveness of American military forces. Krepinevich concluded that, while "new technologies or systems" like those showcased in the Gulf War did suggest a new era in military affairs, the more important development for the future of warfare involved a new approach to "system integration and organizational innovation."[12] As William Perry argued, the capabilities identified as an RMA derived from the larger systems of systems made up of many "links" in a large network of capabilities that, when combined, provided for an increase in effectiveness.[13]

Krepinevich also assessed how new information systems and electronic warfare capabilities were altering the conduct of large-scale armored warfare. At each level of command, the ability to integrate new information quickly and achieve unity of purpose across land, sea, and air forces facilitated a new level of sophistication in command and control. One of the most important contributions of information systems and new intelligence capabilities involved the ability to quickly and accurately leverage real time information for shared situational awareness and the coordination of strikes. Commanders were able to orchestrate large, dynamic engagements with many simultaneous operations. New weapons systems designed for deep battle in Europe provided an order of magnitude increase in the effectiveness of indirect munitions.

Even as Krepinevch completed his study, defense intellectuals were already embracing the term RMA; few recognized the long history of the term in Soviet military thought.[14] Talk of an emergent American-led revolution in military capabilities had first gained prominence in the late 1980s as defense analysts noted important shifts in US military technology. These shifts were labeled a military technical revolution (MTR). MTRs involved the "simultaneous change in military organization and doctrine as a result of technological advances."[15] The term was short-lived, primarily because it focused attention on technology rather than the full range of technological, doctrinal, organizational, and operational innovations associated with changes in US military effectiveness.

An American RMA thesis came to dominate defense modernization discussions. This thesis held that a revolutionary shift in US military power had occurred based on the synergy of advanced ISR capabilities, automated target identification and precision strike systems, information-enabled weapons, stealth aircraft, superior education and training, an offensive military doctrine, and joint warfighting capabilities.

Political scientist James Der Derian noted that the American RMA thesis was "only an idea in the wind" in 1993 when Andrew Marshall circulated an eight-page memo entitled "Some Thoughts on Military Revolutions."[16] By 1994, at least five Pentagon task forces were exploring the notion of an RMA.[17]

Strategic analysts Steven Metz and James Kievit noted that a "heady vision" associated with the evolving RMA thesis "aroused tremendous excitement among American defense planners" by 1995 and that, for many, the RMA's promise of "increased effectiveness at reduced cost" was "an obsession."[18] A year later, military historian Dennis Showalter observed that the term RMA had "replaced TQM [Total Quality Management] as the acronym of choice in the U.S. Armed Forces."[19]

Increased awareness to RMA theory and language in the early 1990s stemmed from studies and conferences, the publication of historical case studies, and theoretical debates about what constituted a definition of significant military change. By mid-decade, references to the American RMA were ubiquitous. Competing definitions of what constituted an RMA dominated much of the discourse.

For Krepinevich, an RMA "occurs when the application of new technologies into a significant number of military systems combines with innovative operational concepts and organizational adaptation in a way that fundamentally alters the character and conduct of conflict. It does so by producing a dramatic increase—often an order of magnitude or greater—in the combat potential and military effectiveness of armed forces."[20]

Paul Van Riper and F. G. Hoffman suggest a more succinct definition, positing that RMAs occur "when a significant discontinuous increase in military capability is created by the innovative interaction of new technologies, operational concepts, and organizational structures."[21] James Fitz-Simonds and Jan van Tol add that "the essence of an RMA" is "not the rapidity of the change in military effectiveness relative to opponents" but "the magnitude of the change compared with preexisting military capabilities."[22]

RAND Corporation defense analyst Richard O. Hundley takes another approach. He observes two characteristics common to RMAs that may occur simultaneously: (1) a new capability "renders obsolete or irrelevant one or more core competencies of a dominant player"; (2) a new operational reality "creates one or more core competencies" involving "some new dimension of warfare."[23]

Military historians Williamson Murray and MacGregor Knox introduced their survey, *The Dynamics of Military Revolution, 1300–2050*, with the observation that the RMA thesis constituted "the heart of debates within the Pentagon over future strategy."[24] Aspects of the RMA thesis appeared in the Secretary of Defense's *Annual Report to the President and Congress*, in a series of new joint warfighting publications, and in Service modernization roadmaps. Public statements by senior military and civilian leaders embraced the central tenets of the RMA thesis.

Quibbling over definitions and a sudden fascination with RMA discourse obscured a larger problem. National security scholar Stephen Biddle argues that the RMA thesis evolved "from exposition to consideration for implementation as a US government policy" so quickly that it "outpaced the ability of scholarship to examine its underlying premises and evidence."[25] A larger problem was the very fascination with RMA theory itself. Although the RMAizing of defense discourse engendered serious consideration of historical shifts in military effectiveness, it focused attention on the concept and manifestation of RMAs rather than serious consideration of military innovation processes that antecede them.

Why the rapid ascent of RMA language? Perhaps the generation of security studies scholars and military theorists that lived through the post-Vietnam renewal welcomed any alternative to the stale concepts and language of nuclear strategy. Or, perhaps a revolutionary lexicon simply seemed more appropriate for an era in which the Berlin Wall was torn down, the Soviet Union dissolved, and nuclear weapons were withdrawn from Europe.

Other factors encouraged the rapid ascent of "revolutionary" references to US military capabilities after the Gulf War, including a focus on the future. A "post-ism" *gestalt* swept through academic and policy communities, with the new world order described as post-Cold War, post-industrial, post-modern, post-positivist, post-nuclear, and post-communist. Swirling in the cognitive landscape during this time, and adding to the sense that a watershed had occurred, were popular discussions of "the end of history"[26] and the irrelevance of Clausewitz's theory of warfare.[27]

Additional revolutionary language came from business and management circles, at the time dominated by business process re-engineering theories that

Defense management proponents fashioned into their own "revolution in business affairs" to compliment the RMA. Change management theorists pushed quick fix solutions that fed on an atmosphere supportive of sudden change. Business process revolution approaches were characterized by a "change or die" philosophy, which became a mantra in boardrooms.

Across government and industry, there was widespread acceptance of the idea of, if not an expectation for, rapid modernization and reform. Drawing on, and later reinforcing, pro-change mindsets were a cadre of reformers seeking to revolutionize government processes and improve responsiveness to citizens.[28] As reformers sought to reinvent government, institute performance measurement, and overhaul financial management practices, defense reformers aimed toward new organizational, fiscal, and administrative efficiencies. A defense reform initiative was launched, base closures and re-alignment commissions recommended infrastructure cuts, intelligence and defense budgets were slashed, and nuclear weapons were withdrawn from most deployed military units.

Toward the end of the decade, prominent public policy scholars voiced concerns about RMA discourse. In 1997, Colin S. Gray called for "scholarly literature expressing deep skepticism about RMA concepts and [information] warfare" to balance the a priori assumptions being made about potential "revolutionary" changes in military affairs.[29] He opined that RMA discussions were in danger of becoming a "Big American Defense Debate" that yielded "more noise than illumination."[30] Defense analyst Michael O'Hanlon expressed similar concerns, concluding that "RMA literature" failed to provide "a systemic assessment of where defense technology [was] headed."[31] His 2000s *Technological Change and the Future of Warfare* opened with a chapter entitled "The So-Called Revolution in Military Affairs" that found programmatic evidence for the RMA at best "inconclusive."[32]

As the RMA thesis ascended in post-Cold War US thought, core arguments from the offset strategy and Follow-on Forces Attack thinking were integrated into descriptions of what new weapons systems meant for the future of strategic warfare. The ultimate expression of the RMA ideal, for many, was the familiar argument that advanced conventional capabilities could reduce reliance on nuclear weapons. Marshall, for example, argued in a 1995 testimony to the Senate Armed Forces Committee that, "Long-range precision strike weapons coupled to very effective sensors and command and control systems will come to dominate much of warfare."[33]

Two years later, a Congressional panel appointed to assess the Quadrennial Defense Review concluded, "Advancing military technologies that merge the capabilities of information systems with precision guided weaponry and real-time targeting and other new weapon systems may provide a supplement or alternative to the nuclear arsenals of the Cold War."[34] And in 1999 Paul Nitze, author of NSC-68, could find "no compelling reason why we should not unilaterally get rid of our nuclear weapons" given our ability to achieve "our objectives with conventional weapons."[35] Andrew Krepinevich and Steven Kosiak added that "potential adversaries would see the US strategic deterrence as more credible if it included forces capable of conducting offensive nonnuclear strategic strike operations."[36]

As RMA terminology and imagery evolved throughout the decade, US defense analysts centered concepts and terms from the digital information revolution into military thought and doctrine. Terms like information superiority, dominant battlespace knowledge, decision superiority, full spectrum dominance, and others displaced Cold War terms and concepts. The lexical turn away from the dominant narrative of nuclear strategy discussed in this study paralleled the thirty-year transformation in US defense strategy and military thought. This thirty-year transformation began in the early 1970s, reached a tipping point in the early 1990s, and culminated in the early 2000s as a new period in US military planning began. Much continuity remained in military thought and operational practice. Still, the narrative of military strategy was fundamentally changed. Nuclear strategy discussions slipped to the margins of defense planning.

Some suggest that the actual utility of nuclear weapons as a military and even diplomatic tool ebbed in the latter decades of the Cold War. As the conclusion to Chapter 3 argued, nuclear strategy and targeting plans failed to provide a useful organizing principle for military doctrine. The focus on nuclear strategy delayed the evolution of operational art. Nuclear diplomacy, arguably, also lost its allure as the Cold War evolved. In a classified study of American security policy, John Hattendorf found that "the presence of nuclear weapons played a role" in diplomacy some nineteen times from 1946 through 1974 but found no cases where they were used explicitly "for political purposes from 1975 to 1984."[37] For some, this reinforced the argument that nuclear weapons had become less important in security affairs; others believed nuclear weapons were a liability.

In the immediate post-Vietnam era, defense planners sought to bolster the credibility of nuclear deterrence by modernizing strategic nuclear systems. By the mid-1970s, conventional force modernization became a key adjunct to doing so. In the early 2000s, strategic *nonnuclear* weapons were perceived as the cornerstone of American military strategy. An interesting reversal occurred at the end of the thirty-year transformation. Defense planners began discussing requirements for new, low-yield, bunker-busting nuclear weapons as an adjunct to America's ability to destroy hardened, underground facilities that terrorists or rogue states could use to develop weapons of mass destruction.

In assessing RMA theory as a guide to understanding strategic history, military effectiveness scholars Williamson Murray and MacGregor Knox conclude that "few works throw light on the concept's past, help situate it or the phenomena it claims to describe within a sophisticated historical framework, or offer much guidance in understanding the potential magnitude and direction of changes in future warfare."[38] Scholarly interest in military effectiveness studies, meanwhile, exposed additional problems of a military revolution discourse that was not primarily focused on the military evolution of the capabilities that won the Gulf War.

Additional criticism came from military leaders. Admiral Bill Owens (retired) served as the Vice Chairman of the Joint Chiefs of Staff from 1994 to 1996 and oversaw the publication of visionary documents such as *Joint Vision 2010*, a seminal document articulating a vision for post-Cold War joint doctrine. Known inside the Pentagon as a forceful advocate for advanced ISR and situational awareness capabilities, Owens frequently testified before Congress in support of sustained funding for military innovation. Although not entirely successful, his efforts helped usher in key concepts like network centric warfare, an integrated vision of how air, land, space, and sea assets could leverage information systems, and the need for additional space communication systems. Among his legacies were a strengthened Joint Requirements Oversight Council (JROC) and a more rigorous Joint Warfighting Capabilities Assessment (JWCA) process.

Reflecting on the pace and scope of modernization, Owens later conceded that the "Pentagon was not really interested in pushing a revolution in military affairs, and few in other parts of the executive branch or in Congress were either."[39] "We changed the vocabulary" and "modified planning processes, established new planning instruments, and adjusted the style, stakes, and procedures in the planning process," he recalls, but "we made less progress than we had hoped" in transforming "the future size, structure, and character of the US military."[40]

Despite the potential for building on previous developments to truly enable the potential of the technologies labeled RMA-like, Owens concludes that the "Pentagon was not interested in embracing" the full promise of the RMA. Owens describes the underlying pathology: although much "rhetoric filled military journals and public pronouncements about 'new eras', 'peace dividends', and 'military revolutions', the U.S. military was quite happy to avoid" significant change while defense leadership merely rode the post-Gulf War "crest of victory."[41]

While the underlying capabilities labeled an RMA retained their revolutionary nature when viewed through the lens of military history, and although defense intellectuals benefited from RMA theory as a framework for thinking about historic changes in military affairs, much of the RMAizing of US defense policy devolved into empty rhetoric. An old wine, new bottles syndrome inhered in many programs as funding cuts led to bureaucratic entrenchment and infighting among the services. Douglas Macgregor, an Army officer that led thinking on the need for a reorganization of post-Cold War ground forces, joined the critics. Reflecting on 1990s defense modernization efforts, he lamented that defense modernization discourse resulted in mere "bumper stickers" that did "not prevent competing service requirements from dominating joint integration efforts."[42]

Successive planning activities and documents record the lack of interest in furthering the spirit of military innovation that had dominated modernization planning in the late 1970s and early 1980s. The 1993 *Report of the Bottom-Up Review*, the 1994 *Nuclear Posture Review*, the 1994 commission on the roles and missions of the armed forces, and the 1997 *Report of the Quadrennial Defense Review* all failed at being "decisive in setting clear guidance" or establishing "a consensus for policy objectives."[43] "In each of these cases," American foreign policy scholar Janne Nolan concludes, "senior leaders, beginning with the president, proved reluctant to engage the issues directly or to provide leadership to guide the outcome."[44]

Major defense planning documents, including the three cited above, drew heavily from Cold War planning assumptions that had evolved over decades based on learning about the Soviet threat. A significant cross-border armored attack remained the chief planning scenario driving force allocation requirements despite the marked decline in the mechanized forces of potential enemies. Planners framed regional instability and political violence as lesser, wholly included scenarios that forces designed for cross-border contingencies could cope with. Rapid dominance and rapid deployment concepts from the Persian Gulf contingency planning were reinforced.

Political and fiscal realities were ill suited to furthering the innovation paths begun in earlier decades. Modernization efforts were delayed, with funding directed to redressing critical readiness issues associated with successive military interventions abroad. To worsen the situation, recruitment and retention levels declined. Services temporarily lowered aptitude standards, accepting category IV recruits (scoring the lowest on tests) at a time when the operational environment grew more complex and weapons systems more complicated to operate.

Department of Defense planning guidance for 1994 and 1995, moreover, anticipated savings of some twenty-five billion dollars from cutting force structure and reducing support infrastructure. The General Accounting Office "blasted these assumptions" in 1996, arguing "there is no significant infrastructure savings."[45] Modernization funding continued to suffer, leading analysts to conclude in 1998 that "defense procurement" was "down by more than 70 percent since its high point in the mid-1980s" and "billions below the requirements to recapitalize America's defense forces."[46]

An absence of urgency for transformation in part reflected the lack of a compelling political reason to embrace it. Voters were not concerned about the level of defense spending. One 1995 survey "revealed that 73 percent of Americans polled believed there were no threats for which the US military was unprepared"; 53 percent believed the US was spending too much on defense; less than 10 percent rated defense as an important issue.[47] Defense issues did not factor into either party's 1996 convention speeches.

Despite a renaissance in American military thought, and perhaps because new thinking was not affecting change evenly across mission areas or quickly enough within Services, Cohen noted "a sense of intellectual and doctrinal stagnation" among some military leaders.[48] For these and other reasons, Cohen posits, emerging transformation discourse represented "more than politics or the quest for novelty": defense reform was in need of an overhaul.[49] Additional concerns surfaced about joint experimentation and operational prototyping, two primary routes for integrating new technology into operations.

For these and other reason, the 1990s are best considered a transition period in which leaders did not or could seize opportunities for meaningful, "transformational," defense reforms. Important to this study are developments in how defense intellectuals used RMA terminology to define defense modernization as they struggled with an uncertain threat landscape.

This does not mean RMA discussions were altogether feckless. As Gray argues, "the raising of the RMA flag mobilized a wide variety of perspectives and skills" and "enabled some long antecedent ideas and streams of analysis to play significantly in a contemporary debate."[50] Indeed, they set the stage for later policy discussions and provided much of the language used to frame a new transformation strategy.

From RMA thesis to transformation policy

The term "transformation" was present within but not central to 1990s defense policy discourse. References to transformation during much of the decade appeared primarily within RMA discussions about historically profound changes in military history or as a descriptor for specific leaps in the effectiveness of one or more combat arms. The term's ascent in defense planning discourse toward the end of the decade was facilitated by a number of widely cited sources, some warranting mention here to gain insight into how defense planning discourse evolved.

Among the earliest and most notable treatments of transformation was Martin Van Creveld's *The Transformation of War* (1991), a book length essay on the changing nature of warfare and its implications for how the armed forces would have to evolve.[51] Appearing on several professional reading lists for US military officers, his thesis that Clausewitz's theories no longer applied became fashionable. His argument that Cold War armies designed to fight other nation states would be useless in the coming era of terrorism and civil wars was prescient.

Among the benchmarks in the shift toward transformation dialogue was the 1997 National Defense Panel report "Transforming Defense—National Security in the 21st Century." Around this time, defense analysts and military theorists began addressing the issue of defense transformation and called for a renewed planning debate.

The National Defense Panel represented an evolution in thinking among defense interlocutors about the pace and scope of defense modernization. Its report argued for initiatives to fully leverage information technology, to develop additional space-based capabilities, and to accelerate organizational innovations.

The 1998 Department of Defense *Annual Report to the President and the Congress*, through which the Secretary of Defense communicates the status of defense readiness and planning, is a second document marking the shift from RMA language to transformation processes.[52] Previous annual reports contained rather pedestrian discussions of RMAs as historical phenomena, using the RMA construct to conceptualize ongoing changes in the nature of war. Of the five main sections of the 1998 report, one addressed Service transformation and another Department-wide transformation. Additional text was devoted to "New Operational Concepts" and "Implementation" needs, including experimentations, demonstrations, and other activities required to facilitate a larger transformation effort.[53]

Subsequent annual reports addressed transformation strategy in lieu of the previous editions' RMA chapters. The switch represented the socialization of "transformation" as a term of art within defense planning and policy circles; for many it suggested that something more prescriptive than existing RMA language was needed.

More forceful transformation language was included in a September 1999 Defense Science Board (DSB) report entitled "DoD Warfighting Transformation." The DSB defined transformation as "a process that seeks fundamental change in how an enterprise conducts its business" in pursuit of "discontinuous change in the nation's capabilities to conduct" military operations.[54] Departing from the deterministic and exogenous view of organizational change inherent in some RMA discussions, transformation was defined as a "self-inflicted" process seeking "very big change."[55]

Among the few muted references to RMAs in the DSB report was the recognition that "very big change" was "sometimes characterized by the term revolution in military affairs."[56] Apparently, rather than accepting the passivity of an "RMA-is-certain" approach, the DSB articulated a view of defense reform and modernization that sought an alternative to RMA-associated rhetoric. Instead of questing after immediate, revolutionary reforms, transformation was characterized as "defining and implementing a vision of the future different from the one embedded, if only implicitly, in DoD's current plans and programs."[57]

It is noteworthy that, unlike studies earlier in the decade, the DSB and other defense planning reports did not begin with assumptions about an ongoing or imminent RMA, choosing instead to focus on transformational processes needed to revamp US defense modernization. An emphasis was placed on military effectiveness. Presumably, had the same report been commissioned even two years earlier, the title and tone of the report would have reflected the centrality of the RMA thesis in official and scholarly thinking about defense modernization.

The DSB, moreover, was among a number of quasi-official defense policy organizations that "did not find much sense of urgency" for significant change in Service warfighting capabilities. The DSB, for example, concluded that "the focus and effort needed" to transform was being underestimated.[58] Despite concerted efforts within the Office of the Secretary of Defense to push modernization, the Clinton administration did not make transformation a priority in terms of leadership attention, a willingness to expend political capital to influence Service decisions, or a clear vision for change conjoined with "sticks" to induce compliance. This left the military services to define their own visions without an overarching mandate to change—or to integrate. Defense transformation did not seem to be getting any traction.

The pace of operations did not help, a fact the Joint Chiefs made clear in September 1998 Congressional testimony by highlighting the inconsistencies between the stated objectives of transformation and the actual resources devoted to transformational programs and processes. The pace of current operations, many lamented, was forcing defense planners to mortgage the future. Because defense budgets did not increase to support the pace of operations, the services were forced to cut new procurements and modernization budgets. The future was mortgaged to pay for current operations, a reality that was also true in the intelligence community.

Complicating matters, the climate of reform and fascination with RMAs and "reengineering" in Washington gave the appearance that "change" was widespread. Champions of the defense status quo actually used the 1991 Gulf War to rationalize the continuation of long-standing organizational structures and weapons systems. There was no shift in the prioritization of technology research and development efforts. This was still the case in September 1999 when then presidential candidate George W. Bush promoted his vision of defense transformation.

The pace and scope of US defense modernization was scrutinized during and after the 2000 election. Writing in 2004, Eliot Cohen recounted the essence of the ensuing criticism. Strategically, American defense strategy was too wedded to "a Cold War-derived understanding of military power" and failed to "focus on the challenges of the new century: homeland defense, a rising China, and what can only be termed 'imperial policing'."[59] Even after a decade of an RMAized defense discourse and numerous visions for future warfighting, technology development and procurement processes adhered to "Cold War paths," leaving "systems suited for a war in Europe with the defunct Soviet Union rather than hardware optimized for" emerging threats.[60] Former defense official Ashton Carter urged the incoming Bush administration to transform defense and to revamp research and development. The United States, he argued, was "not fully exploiting or staying abreast of the information revolution."[61]

Progress remaking the armed forces drew criticism from across the political spectrum. The Army's modernization plan for the twenty-first century, dubbed the "Army Transformation Strategy," was criticized for being more about process and theory than substantive change in force structure. According to Andrew Bacevich, furthermore, although transformation across the Defense Department portended something novel or new, in reality transformation discussions indicated that "the debate over military reform" in the post-Cold War era "had come full circle" back to the early 1990s.[62]

Soon after assuming office in 2001, the George W. Bush administration convened several panels and commissioned numerous studies to chart a new course for US defense modernization. Of note is the reinvigoration of the role assigned to Andrew Marshall and the Office of Net Assessment after the office's marginalization during the Clinton administration.

Marshall was tasked to rethink, re-look, and revitalize efforts to modernize the US military. Ostensibly, Marshall returned to first instincts, in this case the approach taken in the early 1990s when the idea of an MTR (and then RMA) took root among defense planners. Once again, the Office of Net Assessment sponsored panels and studies to explore changes in warfare. These studies benefited from a decade of thinking about changes in how warfare should be waged in the information age. Among the themes were arguments for lighter, more lethal forces able to befuddle opponents with rapid dominance. Many believed that armored forces would not be needed once units were fully enabled by information technology and ever-more advanced knowledge-to-action capabilities. Troops would be wrapped in a protective layer of information dominance.

A Transformation Study Group provided its findings to Secretary of Defense Donald Rumsfeld in late April 2001. The report supported the institutionalization of a new defense transformation process, describing the process as potentially facilitating "changes in the concepts, organization, process, technology application and equipment through which significant gains in operational effectiveness, operating efficiencies and/or cost reductions are achieved."[63] Not ready to implement the report's findings, however, the administration postponed changing defense programs until it could complete a strategic planning process that fully considered the risks involved in such changes. Much uncertainty remained in terms of how fast defense transformation could proceed. By the end of the first George W. Bush administration, nevertheless, "transformation" was adopted as an umbrella term for attempts to remake US armed forces in the model of a lighter, more agile, information-enabled precision force wielding greater lethality over greater distances in less time.

From one perspective, the remaking of post-Cold War defense strategy discussions within a transformation construct revisited some of the same operational challenges and innovation themes addressed in the 1970s and 1980s. In the mid-2000s, the core arguments of the original offset strategy seemed to be undergoing another post-Cold War resurrection. This time the strategic and operational necessity impelling changes in defense strategy and military thought concerned perceived inadequacies to succeed at stabilization missions, including nation building, peace keeping, and counterinsurgency.

No prominent changes occurred during the Bush administration's first year in office. Critics resorted to citing Bush's own campaign speeches lamenting that the American military was "still organized more for Cold War threats than for the challenges of the new century—for industrial-age operations, rather than information age battles."[64] They asked when he would live up to his promise to correct what he called the "the last seven years" of "inertia and idle talk."[65]

Among the planning documents drafted at the time was a new *Quadrennial Defense Review*. It placed transformation at the center of US defense planning. Drawing in part on the above mentioned DSB report on transformation, which called for a transformation cadre to champion reform, an Office of Force Transformation (OFT) was formed to encourage discovery and invention, to help formulate prototyping activities, and to expedite the delivery of new capabilities and technologies to deployed forces. OFT Director Vice Admiral (Retired) Arthur Cebrowski described the objective of transformation as fielding "new sources of power" that "yield profound increases in competitive advantage."[66]

Defense transformation plans were themselves transformed by the September 2001 terrorist attacks on the World Trade Center and the Pentagon and, subsequently, by the global war against terrorism. Indeed, in a November 1 *Washington Post* op-ed Secretary of Defense Donald Rumsfeld referred to the attacks on America as "a wake up call" that created a "new sense of urgency" for modernizing and transforming the armed forces. The bottom line: "Transformation cannot wait."[67] Similarly, then Undersecretary of Defense for Acquisition and Technology Pete Aldridge viewed the war on terrorism as creating "a springboard to transformation" and as stimulating the impetus to overcome the "status quo."[68]

The war on terrorism temporarily accelerated and refined transformation processes. Emphasis was placed on military innovation.[69] The promise of increased defense spending conjoined with this renewed sense of urgency to open a window of opportunity for overcoming cultural, organizational, and philosophical barriers to significant change. Expectations for change increased as a wartime footing provided a context for lowering bureaucratic barriers to innovation.

Transformation assumed the tone of a strategic imperative in the December 2001 Department of Defense *Annual Report to the President and the Congress.* The language of the report implied an accelerated pace and broadened scope, although the administration announced it would delay making significant programmatic changes until all the newly commissioned transformation studies were completed, fully analyzed, and utilized to inform a new defense transformation strategy. Underlying the report was a clear message: the global war on terrorism would not delay transformation.

Discussions of transformation objectives revealed some confusion over the pace and scope of transformation activities. How much change would be attempted at once? As Rumsfeld related in his December 2001 annual report, the Department intended to transform "a portion of the force" to "serve as a vanguard and signal of the changes to come."[70] Among the vanguard models cited was the German experience building a force able to implement the so-called blitzkrieg tactical doctrine of rapid, combined arms mechanized maneuver and attack.

Reawakened interest in military innovation studies refocused attention on factors associated with the rise and diffusion of innovations. Additional thought was given to strategic planning processes and frameworks to manage change. Drawing on business management literature, policy discussions increasingly referenced the need for a mix of innovations, including discontinuous, transformational advances in military effectiveness.

As the defense planning community prepared for a mid-2000s Quadrennial Defense Review, strategy discussions among defense intellectuals focused attention on the need to revamp forces to conduct stabilization missions. Transformation planning, therefore, was shifting from the types of forces that won the 1991 Gulf War and more capable legacy forces that invaded Iraq in 2003 to the forces required to rebuild Iraq and protect the nascent Iraqi government from insurgents and other opposition groups.

The resulting climate for innovation was thus much different in the mid-2000s than it was at the end of the 1990s. DARPA realized a 14 percent funding increase in fiscal year 2002 and an additional 19 percent increase (to 432 million dollars) a year later. During this period, the program responsible for bringing new technologies into operation using Advanced Concepts Technology Demonstrations topped seventy-nine million dollars—a 65 percent increase in funding.[71] The Joint Forces Command changed its focus to participate in many more experiments, looking to find promising new capabilities and practices, rather than focusing on one or two large, overly scripted exercises per year. Defense spending increased, defense policy decisions were firmly in the hands of the Executive Branch, profound shifts were underway in the fabric of national security, and intelligence budgets increased dramatically.

A 2002 Nuclear Posture Review proposed a new, capabilities-based strategic triad integrating new elements of defense, a responsive national infrastructure, and both nuclear *and* nonnuclear strategic strike. This is a critical element in the thirty-year transformation in American military strategy. The new strategic triad is founded on three pillars, only one of which consists of the traditional nuclear bombers, ICBMs, and SLBMs. The other pillars include global conventional strike, missile defenses, and advanced intelligence, planning, and command and control capabilities.[72]

Important areas for advancing military effectiveness emerged during the 1990s as American military forces were tested in successive conflicts against weaker adversaries. Operation Allied Force in Kosovo, for example, engendered greater awareness that additional end-to-end intelligence, surveillance, and reconnaissance capabilities were needed. As information technology facilitated the compression of decision cycles and enriched situational awareness, new operational challenges exposed shortcomings in a key end-to-end process; specifically, the capacities available for persistent battlefield surveillance, the detection and tracking of friendly forces, enemy forces, and noncombatants, the ability to pass timely, geospatially-referenced information including targeting data, precision navigation, and the ability to dynamically retarget weapons, and capabilities for all-weather, immediate post-strike assessments.

The George W. Bush administration empowered a more activist civilian leadership in the Pentagon. Deliberate efforts were undertaken to accelerate and expand the pace and scope of transformation. Secretary Rumsfeld personally over-saw the strategic management of defense modernization efforts. He pressed for new operational constructs and recast the priorities for the future force structure. Civilian leadership challenged the defense establishment to justify programs that did not align with the vision for a smaller, more flexible force. Modernization efforts con-tinued through successive military deployments, including Operation Enduring Freedom (Afghanistan), Operation Iraqi Freedom (OIF), and the more diffuse Global War on Terrorism (GWOT). Experiments and prototyping were encouraged. A formal roadmap process was established to guide modernization efforts.

The emerging era of US defense transformation is building on developments explored in Chapters 4 and 5. New capabilities are being planned, including those incorporating laser weapons, biotechnology, automated global information enter-prises, hydrogen power, the ability to dwell over targets (e.g. endurance UAVs; high-flying airships), combat in space, city-crippling denial of service attacks on critical information services, and robots or semiautonomous "thinking" machines able to self-organize and "swarm." Insurgency warfare in Iraq has also exposed new requirements for urban warfare and national-building, most of which are essentially about more advanced information and decision capabilities.

Revisiting information superiority

The relationship between the discursive (ideational) and material (existential) aspects of the American RMA thesis, including information technology, and early 2000s transformation activities is metaphorically one of a fulcrum and lever. That is, transformation is being considered a way to leverage the RMA, or more accurately technologies and capabilities labeled RMA-like, to transform the Services. The pivot around which this leverage is exerted is a nexus of information technologies and decision support capabilities at the core of significant changes in warfare over the last three decades.

The Bush administration's decision to invade Iraq was premised, in part, on what turned out to be flawed intelligence reporting on Iraqi weapons of mass destruction. This is not the place to debate pre-war intelligence or to question the decision to remove Saddam Hussein from power. Important here is that the invasion decision occurred in the context of a preemption clause being added to national security strategy. Preemption, to be an effective component of national security strategy, requires exquisite intelligence. It requires deep insights into adversary capabilities and intent, accurate indications and warning, prescient decision making capabilities, and superior battlefield intelligence.

Krepinevich noted that the turn toward transformation is in fact a "product of the belief that you are in a period of military revolution. Otherwise, why transform, especially if you're the dominant military."[73] A fundamental shortcoming of both RMA and transformation discussions is failure to appreciate how intelligence informed the military innovation process that produced the revolutionary capabilities underpinning American military dominance.

The generation of American military officers that ascended to brigade and division commands in the 1990s were the first generation of leaders to employ in combat the vastly superior ISR capabilities developed during the Cold War to offset Soviet numerical superiority. They were also the first to understand the importance of a single, simple objective for the reformers who led the post-Vietnam renewal in command and control. Since the early 1970s, the mantra had been: Tell me where I am, where my buddies are, and where the enemy is; if you provide me with this awareness of the battlefield along with precision weapons, I will prevail even if when outnumbered. A unique co-evolutionary process inhered in which an emerging operational cognition among warfighters, a systems view of applying technology among planners, and a leadership vision for joint capabilities all reinforced the attractiveness of information-enabled capabilities to resolve specific battlefield challenges.

Operationally, the post-Cold War Army gained what would be invaluable operational experience in regional conflicts that focused attention on communications, situational awareness, geolocation, and indirect fire needs. Somalia offered a lesson in the stark reality of urban combat against an enemy that could not be distinguished from the civilian population. It was an enemy aided by radical Islamists teaching local militiamen to down American helicopters with rocket propelled grenades, a tactic battle-tested against Russian forces in Afghanistan a decade earlier. Meanwhile, the Air Force employed precision strike capabilities during the 1990s in diverse combat environments. Doing so exposed the inherent limitations of systems and targeting processes conceived for the battlefields of Cold War Europe. Operation Allied Force in Kosovo surfaced apparent short-comings in heavy lift, logistics, and intelligence support to military operations. Failure to find Serbian armor in Kosovo, or to prosecute timely attacks after they were located, revealed gaps in the "kill chain" that questioned the effectiveness of precision air strikes in small wars and urban settings.

Apparent shortcomings in intelligence support to military operations questioned progress achieving information dominance. This was predictable given the lack of funds for intelligence modernization, the tendency to believe that intelligence expertise developed during the Cold War was easily adapted to post-Cold War issues, and to strategic intelligence that devalued long-term analysis. While defense modernization interlocutors enthusiastically embraced "systems of systems" concepts and sought new "sensor to shooter" capabilities, developments in the "I" category of ISR received a different type of attention. Now, in the age of global terrorism, and with growing attention to domestic ISR need for homeland security, the lexical playing field concerning ISR may be leveling.

The information revolution was the real, tangible side of the American RMA, although the underlying operational aspects of decision innovation were not adequately addressed in RMA works or early defense transformation initiatives. As ISR discussions evolve, including many that do not self-identify as ISR discussions (e.g. internet monitoring, chemical and biological agent detection networks, data visualization tools), it is important that students of national security speak of information dominance in its founding terms. This is one route toward the refocusing of defense transformation discussions on essential aspects of military innovation required to sustain American military dominance.

It is instructive to recall the operational origins of RMA terms like information dominance, information superiority, and decision superiority. "Information dominance," for example, was defined in the early 1990s "as a superior understanding of a (potential) adversary's military, political, social, and economic structures, to include their strengths, weaknesses, locations, and degrees of interdependence, while denying an adversary similar information on friendly assets."[74] For Andrew Krepinevich, author of the earliest official assessment of the trends later labeled an RMA, information dominance was not only "relevant to all levels of conflict, from the grand strategic to the tactics," in the ideal situation "it is established in peacetime and sustained in pre-crisis and crisis periods, and in war."[75]

By the 2000s, however, the deeper meaning of information dominance and other terms seemed all but forgotten, with the ideal vision reduced to a Common Operational Picture (COP) or some other military tool for situational awareness and decision making. Discussions of future surveillance and reconnaissance capabilities seem to be converging into a notion of global persistent surveillance, which has become an umbrella term for an ever-diversifying spectrum of ISR requirements and capabilities. The importance of intelligence in actually realizing the omniscience implied in discussions of persistent surveillance remains insufficiently acknowledged.

Underlying decreased understanding of intelligence is the growing fetish with current intelligence or the "CNN effect"; the affinity for real-time reporting and monitoring that used to be the domain of surveillance and reconnaissance. True, benefits abound in having global, persistent surveillance and hoards of monitors "watching" for signs, observables, signatures, precursors, and other events or activities. The pursuit of a more systemic surveillance capability that links air, space, and ground sensors in ways relevant to domestic and foreign security concerns remains a high priority on the national security agenda. Surely, more information about the current operational environment is beneficial. But an important question remains. How are increased information flows translated into increased understanding of the operational challenges that transformation aims to overcome?

There may by a significant potential for diminishing returns on additional ISR investments unless investments are balanced. Colonel Kevin Cunningham, former Dean of the Army War College, concluded that, while it is plausible to expect "that the next generation of technical systems will be that much better at seeing, counting, and reporting," the success of doing so "can breed misconceptions about the proper balance between technical and more manpower intensive intelligence support functions," including "intelligence analysis." "Having to contend with a higher volume of less valuable information," he continues, "actually makes the analytic process less efficient."[76]

One operational goal of the offset strategy discussed in Chapter 4 was to quantitatively expedite, and qualitatively enhance, the knowledge cycle-times that embody operational decision-making process so that NATO forces could defeat Soviet armored attacks as their echelons assembled for movement.

Central to the information-enabled offset strategy were changes in the way leaders viewed the relationship between the cost of speed (how fast end-to-end information services can function) and the value of time (the premium placed on shorter decision-making cycles). When time is of the essence, the high cost of speed is moderated, especially if one treats the cost of systems, infrastructure, tools, and trained analysts as an opportunity cost in the larger realm of operational success. As computerized, digital information technology became a more important arbiter of military effectiveness during the 1980s, military planners began addressing the issue of information and knowledge velocity.

This is a difficult concept for many to grasp. The cost of speed and the value of time, nevertheless, are fundamental aspects of the modern information age and as such are at the heart of recent changes in the art of war. Space and time considerations drove the evolution of the Army's FM 100-5 from Active Defense through two iterations of AirLand Battle. The larger sociological trend has been described as "space-time compression," in which "the time horizons of both private and public decision-making have shrunk, while satellite communications and declining transport costs have made it increasingly possible to spread those decisions immediately over an ever wider and variegated space."[77]

Defense analysts recognized in the 1990s that the "historical limitation" on military capabilities "has been the length of time required to correlate and fuse data from a variety of sources, process it into information and communicate and display that information to intelligence analysts" and then provide actionable information to decision makers.[78] The larger issue for US national security was characterized by Joseph S. Nye and William A. Owens in their influential 1996 *Foreign Affairs* article, "America's Information Edge":

> The one country that can best lead the information revolution will be more powerful than any other. For the foreseeable future, that country is the United States. America has apparent strength in military power and economic production. Yet its more subtle comparative advantage is its ability to collect, process, act upon, and disseminate information, an edge that will almost certainly grow over the next decade.[79]

Strategically, the information edge is "a force multiplier of American diplomacy" in the same fashion that the US "offset strategy" of the 1970s used information to multiply the power of existing conventional forces to offset Soviet numerical advantages in Europe. It is also the essence of the current way of looking at the world and coping with its problems, the spirit of the age in contemporary military thought.

The information edge benefits those able to collect, process, disseminate, and act upon information faster and better than others. Widespread agreement that information technology could indeed multiply the power of existing weapons systems led to an understanding of information as a weapon in its own right.

For Norman C. Davis, the digital revolution was "based primarily on significant technological advances that have increased our ability to collect vast quantities of precise data; to convert that data into intelligible information by removing extraneous 'noise'; to rapidly and accurately transmit this large quantity of information; to convert this information through responsive, flexible processing to near-complete situation awareness; and, at the limit [of this awareness], to allow accurate predictions of the implications of decisions that may be made or actions that may be taken."[80] What Davis describes is an order of magnitude change in the way we collect, aggregate, analyze, store, retrieve, and exploit information. When considered from the perspective of competition in the international system, it is Nye and Owen's information edge.

Major General Robert Scales, Jr (retired), who led the Army's official post-conflict study of the first Gulf War, observes that the war represented a transition "between two epochs: the fading machine age and the newly emerging information age."[81] Herein lies the underlying historical turn on which the RMA thesis drew its strength as an organizing construct for post-Cold War defense modernization discussions.

One of the noteworthy developments during the 1990s was the emergence of information warfare as a distinct subset of war planning and operations at the strategic level. Military education institutions made information warfare part of their curriculums. The National Defense University began offering classes the subject and created a School of Information Warfare and Strategy. In June of 1995, sixteen men and women graduated from the school as the nation's first accredited "infowar" officers. Meanwhile, the Joint Chiefs of Staff created an information warfare directorate. Its special technical operations component would lead highly classified information warfare developments. On the defensive side of the security equation, a National Infrastructure Protection Center assessed and monitored the nation's information, power, and other critical grids. In 1998, Deputy Secretary of Defense John Hamre announced the creation of a Joint Task Force Commander for Computer Network Defense. Information warfare was indeed an area of profound organizational change and intellectual activity.[82]

"To a much greater extent than ever before," Douglas Macgregor observes, military commanders are "technologically positioned to influence action on the battlefield by directing global military resources to the points in time and space ... critical to the campaign's success."[83] Essential elements in doing so include: quickly and accurately visualizing the battlespace; identifying, geopositioning, and characterizing enemy forces; optimizing one's own capabilities to strike the enemy with minimal casualties; efficiently developing campaign plans; and, conducting long-range strikes with more precision, fewer forces, and greater lethality than any time in human history.

Through the 1990s and into the 2000s, expectations and requirements for information gathering, integration, exploitation, and dissemination capabilities expanded. During this time, language and visions associated with the RMA thesis continued to draw on terms, concepts, and capabilities from information technology and its application for knowledge management. Many argued that the rhetoric was outpacing reality, that the promise of information technology was not being fulfilled.

Along the way, national, theater, and tactical information providers all became focused on operational intelligence that was "actionable." The division of labor that underscored learning about the threat environment all but disappeared. This forced the national intelligence community to respond more frequently to quick turn around, crisis-related intelligence requests traditionally in the purview of military intelligence or defense intelligence. These ad hoc, short-term requirements rub against the long-term, historical, collegiate culture of national intelligence agencies. This is especially true in the case of strategic reconnaissance and surveillance, which are increasingly important to decision makers who tend to expect them to be "CNN, always on" information assets.

The real tension is between ad hoc, immediate intelligence needs and longer term studies that facilitate understanding about threat capabilities and intentions. In the coming decade the scarcity issue, the current lack of sufficient collection resources to satisfy demand, will likely wane with the arrival of new air, ground, and space collection systems. The emerging scarcity issue, which is already manifesting itself in the current sensor era, was characterized by Nye in a more recent expansion on the "information edge" thesis. He identified a core problem in the evolving information age. "Attention rather than information," he argues, "becomes the scarce resource, and those who can distinguish valuable signals from white noise gain power." Although Nye is primarily interested in information in general, not information or intelligence to guide defense transformation, he is right to predict that "power in information flows goes to those who can edit and authoritatively validate information, sorting out what is both correct and important."[84]

But as Cunningham lamented, "as sensors evolved from relatively low film-based cameras to near real-time satellite, aircraft, and drone-based radar and image systems, the urge to see and count seems to have overtaken the need to analyze the implications of these observations."[85]

It is important to recognize that ISR advances were not a product of the RMA as much as an artifact of a larger "driver" (to evoke a term of art from the change-management community) in contemporary society: the continuing computer information revolution, which included sociocultural and psychological factors. Key factors included changing perceptions about data and information as resources, new communications processes, and greater understanding of how technological and organizational factors can enhance or impede the cognitive facets of decision making. Current ISR advances reflect, or derive from, the same underlying changes that affected the US strategic threat landscape from the mid-1970s onward.

The Bush doctrine and preemption

Politicians often use college commencement ceremonies to announce policy initiatives or to shape public thinking about an important issue. In retrospect, President George W. Bush attempted both in his address at the United States Military Academy's June 1, 2002 graduation. For sure, it was an attempt to characterize the emerging post-September 11 security environment and to define a new framework for American national security policy. It was also an attempt to shape perceptions of how America and her allies should respond to terrorism, the early twenty-first century's emerging critical security challenge.

With an audience of nearly 1,000 new Army lieutenants perched on folding chairs between glistening, white goalposts of West Point's Michie Stadium, Bush stated "our security will require all Americans to be forward-looking and resolute," declared that the nation must be "ready for preemptive action when necessary," and announced "this nation will act."[86] This was strong language. The address occurred nine months after the September 11, 2001 terrorist attacks and four months after his January 29, 2002 State of the Union address identifying North Korea, Iran, and Iraq as an "Axis of Evil."

For the West Point graduates, much had changed since their arrival at West Point's historic Thayer Gate four years earlier. The graduates' initial years of service, for sure, would depart from the normal routines of new platoon leaders and the tedium of garrison life, training activities, and personnel issues. And the friends and family filling Michie Stadium's seats no doubt pondered where the new lieutenants would spend their first tours, what missions they would lead.

Few predicted that many of the graduates would soon find themselves leading soldiers in an invasion of Iraq.

In hindsight, Bush's West Point address foreshadowed the administration's incorporation of preemption language into national security strategy. Indeed, invoking the term preemption appears to have been a deliberate attempt to shape American views of the international security environment. Bush's speechwriter, Michael Gerson, believed the West Point address would be the most important speech he ever wrote. Veteran journalist Bob Woodward relates in *Plan of Attack*, his account of the Bush administration's decision to invade Iraq, that Gerson worked closely with the president to craft a speech that would instill a sense of urgency in the public's mind. Gerson wanted to convince the American people that their "security interests" and "ideals" were threatened.[87] "The goal," Woodward reports, "was no less than to change the American mind-set the same way it had been changed at the beginning of the Cold War" by other, notable speeches.

Gerson's research in preparation to write the speech included a careful study of Harry S. Truman's 1947 address to Congress requesting funds to support Greece and Turkey in their fight against communist inroads into their domestic political system, which the press dubbed the Truman Doctrine; Gerson also examined John F. Kennedy's 1961 inaugural address. He likely also read other speeches, including British Prime Minister Winston Churchill's 1946 commencement address at Westminster College in Fulton, Missouri, the Iron Curtain address that shaped public perceptions of the post-Second World War security environment.

With President Harry Truman on the dais behind him, Churchill lamented, "from Stettin in the Baltic to Trieste in the Adriatic, an Iron Curtain has descended across the continent."[88] While Churchill's language did not announce new policy, it did spark debate about the post-Second World War world order and began to shape perceptions about America's leadership role in it. Churchill elegantly captured what soon would be common sentiment among American national security planners contemplating the containment of Soviet expansionism. Churchill also characterized the role he believed the United States would assume in the emerging bipolar world order. Because America stood "at the pinnacle of world power," it was burdened with "awe-inspiring accountability to the future" that required global engagement and a new leadership role in international affairs.[89]

Of course, the United States would not bear the burden alone. To both compliment and balance America's emerging leadership role, Churchill supported a strong United Nations Organization to facilitate and sustain a multilateral security regime. A multilateral security regime would prevent any single nation or group of nations from becoming too powerful.

In the immediate post-Second World War era the United States had a monopoly on the atomic bomb. Yet America's real power was political and economic, not military, strength. American acceptance of a post-Second World War multilateral security arrangement reflected hopes for lasting peace as much as pragmatism. Certainly Truman wanted to avoid conflict with the Soviet Union. He also realized that rapid demobilization and conversion from a wartime economy, coupled with the exhaustion of Allied economies, left few options in the short-term other than a multilateral security approach at the dawn of the Cold War.

In contrast, the boldness of the Bush administration has set different expectations for American foreign and defense policy at the beginning of what many believe will be a decades-long fight against global terrorism and extremism. In this struggle, American economic and political strength remain critical. The cornerstone of American national security policy, however, is military dominance, the primary pillar of Bush's grand strategy. Bush's approach is also pragmatic given the administration's vision of acting to prevent the dangers of letting danger gather.

To the Bush administration, it once again appeared that the United States stood at the pinnacle of world power with an awesome responsibility, one that required unilateral leadership. Nowhere in the West Point address did Bush suggest a strong role for multilateral security regimes. The burden of leading the world into a new era would be shouldered by the United States alone, with many Allies reluctantly in tow.

No clever metaphors would be inserted into the West Point speech to define the post-9.11 era. Doing so wasn't Bush's style. He favored plain talk and a tough tone; he preferred plain policy language the "boys from Lubbock" could understand. Bush's 2002 commencement speech was crafted to outline in straightforward language the essence of the administration's approach to dealing with threats to American interests and willingness to use military force as an implement of policy. The United States would henceforth carry the fight to its enemies, to disrupt their planning, and deal with threats before they matured. American power, military and other, would be used to disarm and disable adversaries before they could act.

The day after the West Point speech, Bush's comments were the lead story in both *The New York Times* and *The Washington Post*. Yet months passed before the significance of the speech's preemption language was fully scrutinized and understood. At the time, few considered it an actual declaration of new policy or truly grasped its implications for military strategy, intelligence gathering, or diplomatic relations. Although the speech was widely reported in the mainstream media, attention in the weeks that followed focused more on the creation of a Department of Homeland Security than the new national security strategy.

The historical importance of the West Point speech, and its place in the history of American unilateralism, was revisited after the publication of the Bush administration's September 2002 *National Security Strategy*. The new strategy sent a much different signal than the deterrence and containment threads central to previous national security strategies. It posited that reliance on deterrence would no longer work, especially against the primary threat facing the nation: terrorists and tyrants gaining access to weapons of mass destruction:

> The United States will not use force in all cases to preempt emerging threats, nor should nations use preemption as a pretext for aggression. Yet in an age where the enemies of civilization openly and actively seek the world's most destructive technologies, the United States cannot remain idle while dangers gather.[90]

For many, the implications of the new strategy were not understood until the March 2003 invasion of Iraq, which ostensibly aimed to prevent a resurgent Iraq armed with weapons of mass destruction from upsetting the regional power balance or, worse, from transferring weapons to terrorists.

Subsequent criticisms of Bush national security strategy would continue to point toward errors in assessing the status of Iraqi weapons of mass destruction programs as an indication of the inherent fallibility of a preemption-based strategic outlook. If the intelligence apparatus was faulty or unreliable, the credibility and therefore legitimacy of military actions could be easily questioned or attacked, might set a dangerous international precedent others would use to justify aggression, and would undermine attempts to promote American military actions as benign.

It is noteworthy that the Bush Doctrine was first unveiled during West Point's yearlong bicentennial celebration, a time of reflection on the institution's founding during Thomas Jefferson's presidency. Jefferson signed the Military Peace Establishment Act creating a corps of engineers and a military academy at West Point, New York, on March 16, 1802 during the formative years of American foreign policy. Jefferson was also the founder of the unilateral tradition in American diplomacy and, for some, an exponent of an expansionist approach to security. He extended the United States to the Pacific, sent the Marines to Tripoli to battle the Barbary Pirates, and generally reinforced the Founding Fathers' tendency toward exceptionalism, the belief that America had a unique role in the world and an obligation to improve it.

Many were quick to point out that the United States had always reserved the right to use military force to preempt an imminent threat or to prevent an attack. The primary difference was that until now it was rarely stated. It had never been outright termed a "doctrine." Others questioned the use of terms, arguing that preemptive strikes were traditionally characterized by military action taken against an adversary that is clearly preparing an attack or hostile act. What Bush outlined, on the other hand, was closer to a preventive strike, which involved attacking before a state became an immediate or imminent threat.

Three factors underpinned the inclusion of a declaratory strategy of preemption in the George W. Bush administration's 2002 *National Security Strategy*, all of them related to what is best described as a thirty-year transformation in American military power. The first factor is an unwavering belief that America, founded as a unique nation destined to promote liberty, has a responsibility to shape world affairs and promote conditions supporting the spread of democracy or at least free trade. The Bush administration was willing to employ military force to shape global politics. The second involves resurrection of a long-standing tenet of American national security strategy, one with roots to the Founding Fathers and, in particular, to Jeffersonian foreign policy: a belief that the United States should reserve the right to act alone and sustain sufficient power to pursue unilateral or unipolar policies power if necessary. The third factor related to the thirty-year transformation concerns the administration's belief in American primacy and the nation's unipolar moment: American power should be wielded to shape world events with or without the support of the Allies.

Most discussions of Bush's inclusion of preemption as a declared foreign policy option focus on the first and second factors. Less attention has been given to the exploration of US primacy, and especially its military power, as a percipient or contributing factor in the evolution of the Bush administration's foreign policy agenda and the offensive undertones shaping its national security strategy.

Military capabilities have a subtle yet critical role in the shaping of policy options by erecting expectations about the uses of and influence for military power in global politics. Throughout the 1980s and into the 1990s, American military forces were deployed to further US interests. Building on Cold War developments, the armed forces honed doctrinal precepts based on rapid dominance, precision strike, joint combined arms teams, and other innovations.

As the United States continues its war on terrorism and violent extremism, US defense transformation strategy aims to change "the nature of military competition," further exploit and sustain US military advantages, and "facilitate a culture of change and innovation in order to maintain competitive advantage in the information age."[91] In forging a new path for military transformation, the emerging generation of reformers can learn from the innovators that matured the offset strategy and related innovations into a new American way of war.

Chapter conclusion

Military historian John A. Lynn recently noted that the "study of rapid and radical military change currently enjoys a vogue among historians, social scientists, and even national security types."[92] After surveying studies on military innovation and defense transformation, however, Lynn found the study of military innovation to be "theory-poor."[93] It is important, in this context, that the revolution implied in the RMA thesis concerned the relative change in effectiveness from one historical period to the next. It was not a "revolution" in the sense of sudden, rapid change. The George W. Bush administration's defense transformation objectives are similarly focused on a long-term shift in the fundamental nature of military power that results in sustained American military dominance. The emerging language of defense transformation, therefore, is imbued with a larger sense of where the nation's armed forces are going in terms of effectiveness, how fast, and to what end.

The above chapters explored the decades-long ascent of American conventional warfighting capabilities during the Cold War with emphasis on the immediate post-Vietnam era. It was during the late 1970s and 1980s that programs, doctrine, training, and other aspects of conventional military effectiveness began their ascent to an unprecedented level of dominance over other state-based armed forces.

One of the questions motivating this exploration of the thirty-year transformation is how the rise of American primacy in nonnuclear or conventional military affairs emboldened policy makers to act more decisively with less concern for multilateral alliances or even coalition support. The ascent of a nonnuclear global strategic strike capability and its relationship to revolutionary foreign policy initiatives is an important but underappreciated component of post-September 11 international security affairs.

The task for military planners and strategists advocating military reform is, arguably, to build on past successes and develop appropriate capabilities (technological, organizational, and operational) to meet future threats. An issue not addressed above is whether current transformation planners and military theorists are taking the US military in the right direction. Are US forces being prepared to fight and win future conflicts? What are the right investments to make to prevail in future conflicts?

Ten years after the 1991 Gulf War, in the aftermath of operations Enduring Freedom (Afghanistan) and Iraqi Freedom, there is much talk about a new American way of warfare. Key elements of these "new way of warfare" discussions seem strikingly familiar to the intellectual core of the offset strategy, albeit concepts are refracted through RMA imagery and language. Consider the vision embedded in the 2003 defense transformation strategy: "an enhanced forward deterrent posture through the integration of new combinations of immediately employable forward stationed and deployed forces; globally available reconnaissance, strike, and command and control (C2) assets; information operations capabilities; and rapidly deployable, highly lethal, and sustainable forces that may come from outside a theater of operations."[94] Another area of continuity with the offset strategy is the increasing importance assigned to systems integration across all domains of national security decision making, including the information and intelligence arenas.

Given the similarities with the thirty-year transformation that helped end the Cold War and underwrote American military dominance in the early twenty-first century, it seems prudent for current transformation planners and defense strategists to learn what they can about military innovation in the 1970s and 1980s. During this time, a generation of military strategists and defense planners exploited the emerging computer information revolution and altered the course of military history. Many of the issues and organizational challenges they grappled with, including the diffusion and adoption of critical innovations into the force, will challenge the current generation of innovators.

Figure 6.1 depicts a notional organizational "space" relating different types of change behavior; it also includes a proposed "zone" for innovation studies that specifically aim to inform national security transformation discussions. It is an operational view of the innovation milieu that attempts to deconstruct how a case-specific assessment of military innovation might move from analysis of contextual and organizational factors to a mix of innovation activities. Conceptually, this is one representation of what a reformer or advocate for an innovation might need to consider when leading change.

This process is analytically accomplished by first posing some basic questions about the intent and essence of potential change. First, does the acceptance and diffusion of the innovation require incremental or discontinuous shifts in the

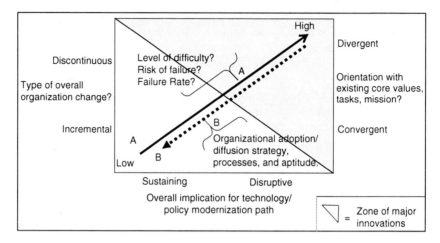

Figure 6.1 An operational view of the innovation milieu.

organization? Second, do the required policy, organizational, technological, or other types of changes lead to the sustaining of current policies or technologies (adapting or extending them) or their disruption?[95] Finally, from an organizational culture and leadership perspective, does change promote a convergence of the old and new or a divergence?

The intent of asking the above three questions is to identify characteristics of innovations within their organizational context as well as thinking about socialization and diffusion capacities needed to achieve adoption. By working through them, one can attempt to locate the innovation along the range of difficulty (line A in Figure 6.1) and then pursue activities (line B) to decrease or mitigate the risk of failure. The intent, over time, is to change the organization so new missions or capabilities are accepted as mainstream or at least are considered as central to the achievement of the mission. They then become part of a new overall capability level from which additional adaptations on the original innovation are pursued. This is a critical part of the innovation diffusion and adoption process. As father of modern economics Alfred Marshall concluded, innovations rarely achieve their full potential until "many minor improvements and subsidiary discoveries have gathered themselves around it."[96]

Based on each particular case, additional analysis is needed to identify what types of risk mitigation and organizational change activities best address reasons why particular innovations fail while others are fully institutionalized. Experiments, prototyping, and incubation activities can be structured and pursued in business units or operational areas most conducive to the particular technological, operational, or organizational innovation. This includes identifying the strategic or operational necessity for the innovation and pursuing diffusion or insertion in ways most conducive to their acceptance. It also requires assessments of what, if any, additional innovations (new capabilities, ideas, operational concepts) are needed to enable the core innovation.

In some cases, convergence involves merely the integration of something old and new; in others it involves an innovative integration or fusion of existing capabilities or technologies. Integration, for example, is central to the story of the Assault Breaker program, which included a joint information fusion element, and to the evolution of the AirLand Battle doctrine, which sought to integrate air and ground capabilities. But the aggregate capabilities represented by the offset strategy represented a divergence from previous capabilities.

Underscoring the above operational view of the innovation milieu is a belief that military innovation is a social process in which technological, operational, and organizational elements conjoin in a specific context. Many approach defense transformation planning from the perspective of technology invention. Research and development to produce new technology, for example, is equated to military innovation. Like other areas of innovation, however, technological innovation is fundamentally a social behavior involving diffusion and adoption processes. For Harvey Brooks, it is "sociotechnical rather than technical."[97]

Regardless of what conceptual frameworks are used to guide defense transformation planning, it is likely that the downside of a decade of RMAized defense strategy will persist. That is, the paucity of sound, policy-relevant historical studies will continue to leave a gap in how decision makers and analysts understand innovation processes and outcomes. Military innovation theories, best practices, and case studies are needed to guide defense transformation.

The thirty-year transformation in American military thought and practice that began in the early 1970s seems to have come to a close in the early 2000s. The bookends of this period are the 1973 settlement leading to the withdrawal of military forces from Vietnam and the 2003 invasion of Iraq. It is important to note that both the beginning and ending of this innovation period are delineated by different approaches to small wars and counterinsurgencies. In the mid-1970s, defense planning efforts largely ignored planning for counterinsurgency warfare. Even had defense planners sought to train, equip, and organize the Army for another small war or for waging counterinsurgency in the 1970s, the political climate did not support doing so. In the 1980s, despite increased attention to small wars and counterterrorism, there was little adaptation in innovation and transformation activities that continued along the path called for in the Offset Strategy.

In the coming years, military innovation scholars will need to provide insights into past cases where military forces have shifted their planning and training efforts to address the operational challenges inherent in small wars and insurgencies. Understanding how the thirty-year transformation arose and how its innovations were adopted into the current American way of warfare is critical to understanding how to best transform organizations, concepts, doctrine, and training to sustain American military dominance across a more diverse spectrum of operations.

7 Conclusion

Revisiting the military innovation framework

In his What is History? E. H. Carr posited that, "Nothing in history is inevitable except in the formal sense that, for it to have happened otherwise, the antecedent causes would have had to be different."[1] True enough. It was certainly not inevitable that late 1970s and early 1980s defense initiatives would culminate in revolutionary shifts in the effectiveness of US conventional warfighting forces as the Cold War ended peacefully. And it was certainly not inevitable that American military forces and defense planners would find themselves struggling with counterinsurgency warfare in the 2000s.

The story of the evolution of American conventional military forces is an understudied and perhaps underappreciated story of military innovation that succeeded. Paradoxically, the thirty-year transformation in military capabilities discussed earlier left American forces ill-prepared for counterinsurgency warfare. As future historians document the history of American military predominance, they will likely find the relatively unknown offset strategy discussed in Chapter 4 to be a critical turning point. Reflecting back on the offset strategy, former Secretary of Defense William Perry relates that his "goal was to offset the Soviet numerical advantage by upgrading American tactical forces with modern technology, with special emphasis on information technology."[2] Perry pushed three priorities: a revolutionary leap in battlefield sensors to locate and identify targets; extremely accurate weapons systems that could rapidly track and destroy a target with a single shot; and the development of stealth aircraft to ensure that American air power could penetrate sophisticated air defenses.

Soviet perceptions of American conventional developments in the late 1970s and early 1980s had an amplifying effect on subsequent US defense modernization decisions. Director of the Office of Net Assessment Andrew Marshall relates that, upon learning of Soviet concerns about "reconnaissance-strike" initiatives in the late 1970s and early 1980s, US defense planners "concluded that it would be useful to intensify those concerns by further investment" in conventional precision strike.[3] "Warsaw Pact defense ministers," Christian Nunlist learned from Soviet archives, "saw developments in conventional armaments in the early 1980s as even more ominous than the strategic change" wrought from nuclear weapons developments because they came with the "revitalization" and "redesigning" of US and NATO conventional doctrine.[4]

From the perspective of the late 1970s, American military thought and defense discourse underwent a near complete revolution by the 1990s. Had they possessed a window into the future, many defense analysts would have found the terms, images, and ideas to be fundamentally different. Terms and concepts developed within the information technology domain were centered within military strategy discourse. By the 2000s, terms such as network centric warfare, information dominance, and other terms would dominate military thought.

Of course, it is impossible to disassociate the invention of the computer, the impact of nuclear weapons, early satellite navigation systems, the advent of radar, or other antecedent factors from the chain of events leading to the emergence of "revolutionary" American military capabilities. That said, and historical contingency aside, it is possible to delimit the period in which decisions were made to fund specific programs and develop certain capabilities that, in time, gave rise to forces exhibiting a discontinuous increase in military effectiveness. Key decisions, inflection points, cognitive and doctrinal turnabouts, technological developments, and innovation activities cohered to create capabilities that altered calculations of strategic effectiveness and how military organizations measured their readiness. As the American defense community turns its attention to transformation, previous periods of military innovation should be scrutinized.

Allan Millet cogently defines the utility of military innovation studies for policy makers. "Knowing how and why innovation flourished or lagged," he contends, "is an essential step toward understanding the enduring dynamics of military innovation and the challenges of military reform."[5] Similarly, political scientist Stephen Peter Rosen intended his *Winning the Next War: Innovation and the Modern Military* to inform defense transformation, with specific implications "about the role of resources, intelligence, and civilian control in military innovation."[6]

Barry Watts and Williamson Murray have argued that the fundamental motivation for studying military innovation is to help defense planners "think creatively about changes in the nature of war" as they attempt to predict what capabilities will secure future battlefield victories; innovation studies should not be undertaken merely to study history for its own sake.[7] Understanding the ebbs and flows of previous innovations, Murray argues elsewhere, illuminates how defense organizations pursue innovation which for contemporary policy makers suggests how innovation and transformation initiatives might alter "performance on the battlefields of the twenty-first century."[8]

Military innovation studies, consisting largely of historical case studies organized around specific theoretical frameworks, provide policy makers and analysts with insights into past innovation processes and outcomes. As Chapter 2 discussed, students of military innovations describe and analyze the conditions common to successful innovation processes to suggest how others might replicate them. Although certainly not sufficient, such innovation processes and outcomes are necessary antecedents to successful military transformations.

Important differences distinguish most RMA works from military innovation studies. With some exceptions, the 1990s RMA debate among defense analysts focused on grand changes in warfare, on technologies likely to dominate twenty-first century conflicts, and on whether or not emerging capabilities deserved the label "revolutionary." Conversely, military innovation studies tend to start with grand challenges to strategy (or smaller ones to tactics) and then relate how organizations overcame them in ways that significantly changed a military force's ability to fight and win in combat.

Research objectives guiding military innovation works vary: descriptive, prescriptive, or a mix of both. Findings and conclusions are increasingly surfaced in policy discourse, including public policy journals, official reports, and dialogue among policy makers themselves. This chapter reviews key aspects of the thirty-year transformation in American military effectiveness, focusing on the foundational period of the new American way of war, through the lens of the innovation framework introduced in Chapter 2. The framework is reprinted as Figure 7.1.

Regardless of the historical data, theories, or methodological bent pursued, military innovation studies seeking to inform policy must provide policy makers with some framework consisting of structural interfaces, insights into human

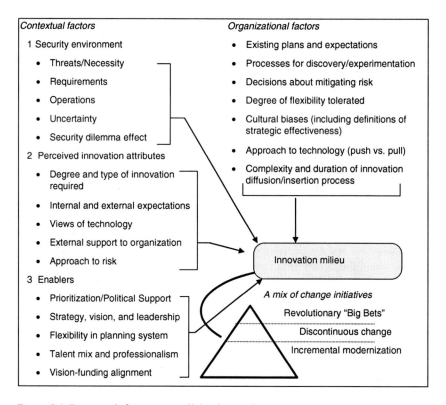

Figure 7.1 Framework for conceptualizing innovation.

behavior, and contextual factors influencing both. Numerous definitions and typologies of innovation exist. I am persuaded that for defense analysts interested in discontinuous changes in military effectiveness, a straightforward innovation framework can be employed to inform analysis.

There are abundant reasons for knowing one's own history before setting out to change it. Policy recommendations should be based on realistic assessments of the future strategic and operational environment, assessments that are only meaningful when placed in historical context. A window appears to be opening for security studies scholars to provide case studies and theoretical tools to policy makers interested in understanding military innovation management and the characteristics associated with planning increases in military effectiveness.

Revisiting context, the security environment, and necessity

Underlying the above innovation framework is a belief that context is the key to understanding innovation behavior and outcomes. The framework suggests one way to consider the interaction effects of the full complement of influences on military innovations, their diffusion and adoption, and their effect on military effectiveness. This includes the primary elements of innovation systems, processes, and actors that exist in specific moments within specific organizational settings.

Contextual factors define the general boundaries and the inherent potential of an innovation milieu. Understanding contextual elements of the larger social, technological, economic, and political environment requires understanding how decisions, organizational processes, key events, and ingrained behavior influence the evolution of defense strategy. This includes how modernization agendas are set by institutions and people. Principle contextual elements include threats (capabilities and intentions), associated requirements emerging from analysis of the gaps between threats and available means to meet them, the nature of operations envisioned in future battles, and the approach taken to managing uncertainty.

Interpretations of the current and future security environment condition individual and organizational understandings of necessity, which, in turn, construct a larger contextual understanding of the basic considerations addressed in national security documents and defense planning. Necessity, an ancient concept in preparations for warfare from Homer to Thucydides to Gibbon to the present, fuels military innovation. It derives from challenges and emboldens opportunity. Cold War developments linked to necessity included the Polaris submarine launched ballistic missile, spy satellites, stealth aircraft, deep strike doctrine, and DARPA's realignment to focus on conventional warfare. Each was also concerned with managing uncertainty in terms of mitigating against the risk of a Soviet attack.

Of course, understanding the effect of necessity on defense planning is much easier with the benefit of hindsight. Organizations, or more accurately individuals and cohorts responsible for decisions within them, do not always accurately identify or characterize the essence of a strategic or operational necessity.

Necessity need not be extant at the time of the innovation in the form of an immediate threat or challenge. Analysts may perceive a decline in capabilities or what management theorists call an anticipated burning bridge. That is, developments in foreign militaries or a shift in the strategic environment that renders one's capabilities less relevant or effective. Examples from Chapters 4 and 5 included accurate Soviet surface-to-air missiles and antiaircraft radar that threatened to weaken the effectiveness of airpower. This impelled the development of radiation-seeking missiles and stealth aircraft to penetrate air defenses and attack command and control sites, air defense radars, and other targets. Discussions of asymmetric counters to current American military predominance are giving rise to a similar burning bridge mentality, but focusing the resulting anxiety on potential battlefield solutions is difficult given the inability to identify specific threats from which to design effective counters.

Important contributions to US understanding of the Soviet threat included Defense Science Board studies, intelligence analysis, and the industry study led by Joe Braddock, which provided a detailed argument for defeating Soviet armored echelons. Of course, intelligence reports often overestimated Soviet military prowess. Still, intelligence analysis of Soviet potential military capabilities did seem to accurately portray Soviet plans, potential battlefield performance, and weaknesses the United States could exploit with long-range precision strike and other initiatives.

Assessments of enemy capabilities were most helpful to technology and doctrinal innovation when they provided specific insights into tactical challenges and potential enemy weaknesses. Solutions to operational challenges are less difficult to implement when they are communicated along with a clear vision that resonates with the troops and when they are guided by rigorous strategic planning and training processes. The components of the long-range precision strike enterprise proposed and demonstrated by Assault Breaker were based on a specific tactical problem from which requirements and a new approach to operations followed. Delay in achieving the underlying vision reflected organizational and political challenges, not technological ones. Planning and training processes must include realistic exercises and be underwritten by a belief that much can be learned from failure. Learning from failure, indeed, remains a principle theme at the National Training Center.

Today, processes for demonstrating a new capability are embodied in prototyping activities, experiments, and advanced concept technology demonstrations. The fundamental objective is sustaining warfighter advocacy based on their ownership of the driving operational needs, ownership reinforced up the chain of command (and accountability) by a commitment to operationally testing new capabilities designed to redress strategic requirements. Because military operations are no longer a single-service activity, constituencies within each military organization must be included in concept demonstrations and the evaluation of their outcomes. Today's innovators should follow the example pursued by Perry and DARPA technologists, who cultivated relationships with champions for experiments and built partnerships with officers that could help translate battlefield requirements into operational prototypes.

This brings us to another aspect of the security environment: requirements. Requirements are operational capabilities, expressed as needs, which organizations deem critical for success on the battlefield. They can include logistics support requirements, intelligence or information needs for decision making or even capability descriptions to guide technology research. They should also provide some links to doctrine, either by referencing existing doctrine or pointing toward needed changes to it.

Requirements gathering, aggregation into capabilities areas, and translation into new research and development initiatives work best when operators define needs based on specific knowledge of the enemy or a specific tactical problem. Innovations, moreover, frequently evolve from a recognition that requirements cannot be met with current or projected capabilities.

Many of the innovations anteceding the American RMA aimed to satisfy specific requirements related to a relatively narrow, yet related, set of strategic and operational threats. In an environment of nuclear parity, worsening East–West relations, and pressure at home and abroad to raise the nuclear threshold, stealth technology, precision strike, maneuver doctrine, and closer cooperation between air and ground forces offered operational solutions to strategic and operational problems in the European theater. Information technology promised to solve command and control, navigation, and coordination challenges. Information superiority presented options for managing uncertainty and facilitated steps toward cross-Services integration without threatening organizational autonomy.

Adoption of doctrinal innovations is made easier by the demonstration of more efficient or more optimal capabilities that give rise to increased military effectiveness. But demonstrations are not enough. New capabilities must be communicated back to warfighters and defense planners in terms of the operational needs and requirements that spawned them. They must be socialized by advocates of the new approach who are recognized experts by the community being asked to adopt the innovation. Another route to successful adoption is identifying the innovation with important new missions or core competencies that an organization is struggling to master.

This is harder to do in a planning environment that is not guided by specific threats. Threat-based defense transformation is often easier than capabilities-based transformation—the current American approach to defense transformation. Some locate the change to a capabilities-based approach to requirements generation in the post-Cold War demise of the Soviet threat. A capabilities approach, however, actually emerged from discussions in the 1980s.

For example, capabilities-based transformation was at the core of NATO's planning for Follow On Forces Attack by extending the battlefield. Although the framework was Soviet military power, planners in the mid-1980s were already developing an innovation model based on mission capability packages that would enable increased military effectiveness across a range of missions based on an underlying operational need or concept. A 1986 US Office of Technology Assessment report, *Technologies for NATO's Follow-On Forces Attack Concept*, argued "systems should be considered not individually, but as complete packages to support specific operational concepts."[9] Not procuring or adequately integrating into forces any one of the required subelements "could greatly reduce the value of investments in the others."[10]

Then Deputy Director of Defense Research and Engineering Frank Kendall, furthermore, argued in 1992 that FOFA evolved into a Joint Precision Interdiction capability, "taking the emphasis off of the non-existent Warsaw Pact threat and placing it on multiple theaters and on critical military targets, including targets at greater operational depths."[11] Critical aspects included an integrated sensor system, precision geolocation and visualization tools, and near-time intelligence reporting.

Uncertainty about the future is a fundamental component of the security environment. The central question is one of risk: what are the consequences of incorrectly identifying a strategic or operational need or the future requirement? The degree of uncertainty about adversary capabilities and intentions determines the degree and nature of risk inherent in choosing one course of development over another. Frequently, as was the case for advanced surveillance, targeting, and strike capabilities, the need for innovation often derives from a realization that existing capabilities cannot guarantee success. Making informed decisions about risk, it seems, requires some definition of what it means to succeed or fail on the battlefield; this reinforces the need to firmly grasp the contextual nature of threats and necessity.

Uncertainty falls into two categories. There are so-called known unknowns, the possible futures one has already identified as potential scenarios from which planning proceeds. Assessing these provides a probability estimate of each scenario actually occurring, along with ideas about how to mitigate the downside of their occurring. More difficult to get right is the risk category that theorists term residual uncertainty: possible futures that are unknown unknowns. Operational or tactical surprise is often avoidable even if one has a solid reading of the known uncertainties. Strategic and technological surprise, which frequently compounds operational surprise, is more likely when organizations have not identified the full range of likely futures. They are more likely when there is a high level of residual uncertainty, which is often caused by a failure in imagination when considering the range of operational capabilities needed to offset future threats.

Strategic surprise leading to national capitulation is the most dangerous and least likely form of uncertainty. A more likely threat to US security involves technological and operational surprise, which are also most frequent types of surprise punctuating military history. Technological and operational surprise can be devastating, militarily and politically, at the regional level. Their occurrence often reflects faltering threat perception or bad decision making. Even if threats are accurately perceived, organizations often make wrong investment decisions even after an operational requirement is documented.

Perceptions of innovation attributes

Another category of contextual factors involved in innovation are perceptions of innovation attributes, the essential aspects of a technological, doctrinal, or organizational change that innovators associate with increased effectiveness. Perceptions are shaped by assessments of the security environment as well as existing beliefs, biases, and prevailing views of military capabilities. How leaders discuss innovations, position them on meeting agendas, and associate them with strategic objectives all influence how others in the organization react to new approaches to solving operational challenges.

Military leaders think about and train for battle by first understanding the security environment and its essential threat characteristics; then they adjust warfighting concepts and tactics to meet them. Social scientists would call the underlying process an operationalization of a theory of warfare or warfighting—the cognitive ordering of variables involved in combat that are causally related to images of or beliefs about how success is actually achieved. Innovations leading to significant increases in military effectiveness are partly derived from a similar process. That is, they involve a realization that some previously unknown variable or capability, thrown into the mix, exposes another way to alter the outcome of a battle.

For this reason, important elements in changing perceptions of military effectiveness are the efficacy, strategic viability, and operational reliability of research and development activities that "demonstrate" the viability of innovations. It is not enough to develop technology or rethink doctrine. As mentioned above, they must be tested, proven in realistic exercises, debated, and competed against the status quo. Without an opportunity to prove their potential, planners are hard-pressed to recommend a new, potentially risky program that disrupts existing, incremental development work.

Among the prominent undercurrents from previous chapters were the adoption of a systems approach to warfighting and a capabilities approach to force modernization. The former spawned operational art, the latter an approach to defense planning that gradually built on the idea of mission needs statements and the total cost of mission execution.

Expectations, restating the point, are important. Expectations are fundamentally constructed on perceptions, attitudes, and the cognitive lenses through which information about the world is interpreted. From the perspective of understanding innovation behavior, the construction of these perceptions and expectations forms an important part of the context in which "change" initiatives are embedded. Especially important are how an organization goes about communicating expectations as uncertainty about the operational objectives for an innovation, uncertainly about how the innovation will or will not help an organization suceed in the future can undermine innovation activities.

Expectations about future capabilities cascade down and through organizations, conditioning views of behavior, performance, and all manner of priorities—from procurement, to research and development, to training and even recruitment. All of this influences how different communities of practice, either operational or applied science, initially develop options for the degree and types of innovation required to address threats, bridge gaps in capabilities, and invest in new capabilities to reduce risk. Expectations also represent the bias civilian and military leaders have about theories for achieving victory. They are not easy to change.

During most of the Cold War, for example, the US military trained and equipped for scenarios involving Soviet attacks into NATO territory along several mobility corridors. They also planned for a North Korean attack into South Korea. Public servants and military officers spent entire careers planning for a handful of scenarios, each of which had a highly evolved theory for victory. The new capabilities pursued in the 1970s grew out of fears that Soviet precision strike capabilities, some which were demonstrated during the October War, had shifted the balance of power in Europe decidedly against NATO forces. Partly in response to new Soviet military advances, long-range precision strike capabilities were integrated into US forces to increase NATO's ability to defend against an armored advance. These new American capabilities initially gained acceptance because they aimed to replicate the effects of small nuclear weapons on Soviet armored echelons, reinforcing ingrained theories of victory.

Despite barriers to acceptance, AirLand Battle and joint operations in support of Deep Attack gained acceptance. In the early 1980s, programs like Assault Breaker and the coordination of Stealth air strikes were developed specifically to alter the calculus of success. By the end of the 1980s, a fairly well defined process for technology push and pull ensured that certain types of new technologies were at least known to the respective services and considered in modernization planning. In addition to pulling ideas, technology, and operational concepts into defense reform discussions, the core set of strategic and operational threats driving European security assessments in the 1970s and 1980s provided an organizing framework to channel additional forms of organizational creativity.

The above innovation framework depicts both internal and external expectations as factors to be included in the contextual domain of perceived innovation attributes. They are elements of the cognitive context. Innovations are nested within organizations that are themselves part of larger organizations that exist in still larger social–political contexts, and so on. Internal and external expectations are among the most important for military innovation scholars to understand. Difficult to empirically document, expectations nonetheless play a large role in how innovations diffuse within and across organizations.

Internal expectations apply to organizations that will actually implement the innovation; external expectations apply to those organizations, leaders, and other entities for which support is required to successfully diffuse the innovation. Expectations about a potential NATO–Warsaw Pact war were for decades driven by a belief that any future war would involve nuclear weapons. Conventional operations on nuclear battlefields were expected to be minor in scale, with NATO having to resort to theater nuclear weapons early in the conflict. The Air Force corporately held very low expectations for long-range conventional precision strike or, for that matter, for precision munitions.

Much of this changed in the late 1970s and early 1980s. New exercise and training initiatives, some linked to concept demonstrations and prototyping activities—Assault Breaker was one of them—helped leaders and decision makers visualize the potential of new capabilities. Along the way, the Army refocused its efforts on indirect fire missions, maneuver, and expediting the flow of data from theater surveillance assets to decision makers.

Long-range precision strike emerged as a competency the Army had to master to succeed at its larger ground combat mission. Some concepts called for delivering thousands of precision munitions in the opening battles of any future conflict. Succeeding at ground attack missions required theater intelligence capabilities able to "see" forward some 300 kilometers into enemy territory. This required reliance on national technical means (spy satellites) to assess an adversary's military capabilities, to monitor any changes in readiness suggesting preparations for war, to develop target reference points, and to generate maps for use by forces in any future conflict. Airborne capabilities were also needed to provide local commanders with a more responsive intelligence capability, to deliver targeting data within such shorter timelines, and to monitor the direction, speed, and size of enemy forces moving toward NATO front lines.

During the Cold War, a spectrum of ISR systems evolved to improve national security decision making by also enhancing strategic military capabilities. From the earliest days of the American RMA, end-to-end intelligence, surveillance, and reconnaissance (ISR) capabilities were a priority and viewed as *the* fundamental enabler of success. Over time, especially after the 1981 Polish crisis, an imperative was placed on better warning and crisis monitoring capabilities. Intelligence co-evolved with military capabilities, gradually overcoming traditional bias as a secondary factor in operations. By the end of the Cold War, thanks in large part to the integration of digital information technology into intelligence processes, intelligence support activities began to overcome criticism that its ability to inform battlefield decision making always lagged behind doctrine and operational concepts.

Military innovation scholars will need to adopt a more balanced approach to understanding the origins and evolution of current ISR capabilities if they hope to inform defense policy in the 2000s. This requires greater attention to innovation in intelligence operations, processes, and policies to assure that military *and* civilian, foreign *and* domestic ISR requirements are understood and met. In the emerging round of national security transformation decisions and funding prioritizations, the tendency will undoubtedly be to continue the trend found in defense reform decisions—to pursue advances in surveillance and monitoring capabilities (e.g. the drive for "persistence surveillance") without overhauling or bolstering the intelligence functions that turn such capabilities into a strategic advantage. Additional sensing does not equate to more insight. Particular attention must be paid to leveraging the most important strategic asset in the ISR domain—the analysts, a truism brought into stark relief during the Congressional hearings on intelligence leading up to the September 11, 2001 terrorist attacks.

Capabilities showcased by offset strategy programs like the Assault Breaker technology demonstration were well-known and, by the mid-1980s, generally accepted by defense planners as part of the future force structure. Delays in developing and fielding technology, doctrinal changes, and organizational innovations occurred. But the very existence of innovative programs, along with the knowledge of successful experiments, did reinforce perceptions and expectations across the spectrum of thought leaders in positions that could influence funding decisions. Society was also changing in terms of acceptance of information technology.

Some programs were not championed by military leaders. GPS and JSTARS were essentially "pushed" after successful operational testing in Europe convinced senior military officers that the technology was sorely needed. Unmanned aerial vehicles and systems that proved critical to operational successes in the 2000s were marginalized by defense planners. No matter how hard innovators push solutions to operational problems, adoption will not transpire unless some degree of pull can be engendered among influential users. For this reason, civilian leaders that have the power to overcome and override the intransigence of military services must sometimes plant and cultivate the operational need for a capability among a user base with sway among key influential users. Multiple sources of invention and innovation exist in any modernization process.

Three turning points in expectations about the type of military capabilities required for the evolving security environment occurred in the late 1970s. First, strategic and operational requirements for theater nuclear targeting led to the Presidential Directive 59 in July 1980, which called for a range of capabilities supporting dynamic nuclear targeting, including secure global communications and in-route retargeting. Further refinements in precision location, dynamic targeting and retargeting, and the tighter coupling of command, control, communications, intelligence, surveillance, and reconnaissance suggested new opportunities for nonnuclear theater strike.

A second shift involved Soviet force structure and doctrinal changes that raised concerns about increased operational tempos and the spillover of superpower competition into peripheral regions—most importantly the Persian Gulf. Expectations for campaign planning evolved as domestic and international pressure mounted to raise the nuclear threshold. The rapid deployment approach to regional conflict encouraged additional changes in US force structure, support for nonnuclear long-range strike, and the diversification of associated planning, training, and doctrine. Military responses to aggression required lethal strikes with smaller forces wielding weapon systems capable of greater precision.

A final shift in expectations involved warfare itself, or more accurately new definitions of success. In the early 1980s, planners considered possibilities for prevailing in a future European conflict as well as reversing nearly decades of reluctance to engage militarily abroad. Partly this reflected responses to the evolving Soviet threat and the recognition that military capabilities were needed for new regional missions. The attack on Libya, the first combat use of precision munitions since Vietnam and the prototype counterattack in the current global war against terrorism, was the culmination of years of shifting expectations about military force. These changes, among others discussed in Chapters 4 and 5, involved changing views of technology, especially the value of information technology.

Enablers

Another aspect of context factoring into military innovation studies are enablers. Other names for them include catalysts, facilitators, or influence paths. From one perspective, they are resources that innovators leverage to change the course of modernization. From another, they represent the linkages between the security environment, perceptions of that environment, and specific organizational factors or influences on innovation decisions.

Innovation "enablers" are frequently considered financial in nature, with other types of enablers ignored. But the resources that enable innovation should not be limited to fiscal concerns. Vision and leadership, the mix of available talent, the prioritization of development initiatives, and the paths through which one can change views of technology are all "resources" from the perspective of their collective ability to influence the innovation milieu. We must also remember that strategic communication competencies and the ability to engender cultural change are the most important enablers that managers can use for true "organizational change." The essence of strategic communication is of course different for wartime and peacetime innovation. Getting strategy right is critical to both, of course, as is an understanding of the audiences involved. "With the offset strategy as a guide," for example, William Perry worked to focus "the attention and support of high-level DoD decision makers, Service chiefs and Congress to speed several important technologies from concept to implementation."[12] Visions of future warfighting thereafter built on the key thrusts of situational awareness capabilities for intelligence gathering, target identification, navigation, precision strike, and expedited logistics.

William Perry's support for innovative approaches to technology continued throughout his government service. Later, as Secretary of Defense in the first Clinton Administration, Perry would team with Director of Central Intelligence John Deutch, and Vice Chairman of the Joint Chiefs of Staff Vice Admiral William Owens to form the National Imagery and Mapping Agency (NIMA) in 1986 (renamed the National Geospatial-Intelligence Agency, NGA, in 2003).

A controversial decision at the time, NGA's successful integration of national imagery intelligence capabilities with defense mapping and charting services provided many of the crucial targeting, navigation, and precision strike innovations demonstrated in operations Enduring Freedom and Iraqi Freedom. In some ways, NGA's integration of information technologies and analytic expertise to provide geospatial intelligence represents the evolution of core aspects of the offset strategy and the realization of Perry's precision strike vision on a global basis.

Vision and leadership were also important enablers for the Army in the years immediately following Vietnam. Army and Air Force leaders agreed on the vision of AirLand Battle, expending organizational capital in the process. Vision was provided by the most senior leaders down to combat leaders in the field pushing innovation at the tactical level. The Reagan administration's defense buildup and National Security Decision Directives to bankrupt the Soviet Union reflected the president's "belief that the Cold War was not a set of problems to be fixed, but a situation to be ended."[13] Reagan's vision solidified into policies that prepared the road to Reykjavik and helped bring the Cold War to a peaceful resolution.

An earlier example of vision impelling military innovation was the offset strategy and its range of initiatives. Assault Breaker, with its integration of intelligence, targeting, information dissemination, weapons platforms, command and control, and munitions systems was a defining "force package" of the era, one that encouraged further thinking about long-range precision strike. Most of the weapons systems developed since the 1970s have relied on some information "brains" to work, a continuation of the underlying strategy.

Programs are traditionally managed by balancing adherence to the schedule coordinating subcomponent delivery, integration, and testing, the performance characteristics of the overall system or platform, and the overall system or program cost—including initial transition to service. Risk assessments are performed during the process to identify and prevent schedule slips, degradations of performance, and the myriad exigencies leading to cost overruns. Rarely are systems on time, within budget, and as capable as initially specified. Frequently, one of the three program management elements is considered more important. A particular performance threshold, for example, may be critical, with additional funding and time provided to overcome technological or systems integration challenges. Cost constrained programs, on the other hand, tend to focus on the bottom line rather than maturing capabilities or meeting a specific deadline. They stabilize cost by shaving performance parameters or extending the schedule so costs are addressed over more fiscal planning years.

Significant innovations aiming to fundamentally alter the effectiveness of military organizations usually require a more flexible approach to the balancing of cost, schedule, and performance. In wartime, schedule is usually the most important, with cost less of a concern if the innovation has strategic importance. In other times, cost is considered the most important, especially when the battlefield effectiveness of a significant innovation is relatively uncertain. Flexibility to change the schedule, cost, performance parameters of a program is another organizational enabler of innovation.

All of this does not mean that financial resources are not important. Funding alignment is a key indicator of what organizations consider both important and what leaders think possible in terms of changing the calculus of military effectiveness. In the early 1970s, for example, the allocation of defense dollars to new strategic nuclear systems indicated not only what defense planners considered a key requirement for national security but also what Secretary of Defense Melvin Laird considered possible given the political situation.

Increased defense spending at the end of the decade was a key contextual enabler for DARPA activities. Indeed, DARPA's budget nearly doubled from 1977 through 1981. DARPA realigned its activities to solve operational challenges posed by Soviet conventional forces in large part because of shifts in the security environment, greater willingness to support advanced technology development, visions for how technology could be applied, the empowerment of leaders with specific agendas, and recognition that funding needed to be aligned with strategic objectives for raising the nuclear threshold.

Talent mix is another aspect of the contextual environment military innovation scholars and defense reformers should consider in their approach to transformation. Organizations cannot develop and diffuse significant innovations without some measure of diversity in its talent base. Internal and external expectations, discussed above, influence the evolution of skill sets within societies and security regimes, creating guild-like cohorts whose self worth and value are directly related to views of current and future technology, the degree and type of innovations perceived as beneficial, and approaches to risk. An important development during the maturation period of the American RMA was the rise of information technologies on the margins of traditional military occupation specialties and, over time, the migration of almost every occupation specialty into the information technology domain.

Increased professionalism, a factor related to talent mix, underwrote the American RMA. Lieutenant General (LTG) Stan McChrystal was the Deputy Director of Operations, Joint Chiefs of Staff during Operation Iraqi Freedom and assumed Command of the Joint Special Operations Command in October 2003. He contends that the return of professionalism to the Army—indeed to all of the Services, in the late 1970s and through the 1980s is the most important antecedent to what observers dubbed an RMA in the 1990s. Arguing that the true revolution was one in training and education, LTG McChrystal concludes that any leap in strategic effectiveness associated with American forces at the end of the Cold War derived from a culture valuing learning and the development of leaders. American troops now demonstrate a penchant for innovation in the field—the institutionalization of innovation. The ability to harness technology, using it to offset strategic and operational challenges, to innovate organizationally and operationally: these are the foundations of the American RMA for McChrystal, hallmarks of a modern professional force.

Where LTG McChrystal correlates military professionalism with innovation, retired Admiral Bill Owens sees professionalism as "synonymous with military effectiveness."[14] Innovation in the planning for and conduct of warfare is in fact a key enabler of increased effectiveness. After decades of relative stagnation, a return to a culture of innovation occurred in the late 1970s.[15]

Another dimension of talent mix is the breadth of skills within an organization folded into the innovation consolidation and diffusion process. Beginning in the late 1970s, significant military innovations involved systems integration, spawning several lines of planning and operational processes on which current transformation activities rest. Systems engineering, discussed below, emerged as a skill to facilitate intra-organizational planning by helping organizations develop requirements and plans based on externally driven requirements.

Flexibility in the planning process is critical. Leaders actively promoted innovation in the late 1970s to a much greater degree of self-determination than their immediate predecessors. They fostered greater appreciation for the value of doing things differently and, as the Soviet threat became politically more pronounced, they accepted a greater degree of flexibility in designing solutions to operational challenges.

Things were broken, they needed fixing; or at least this is what a new generation of military and civilian leaders believed. They shared a larger vision for offsetting Soviet advantages without relying on nuclear weapons. More importantly, they realized a critical need for organizational renewal and concerted efforts to instill pride, confidence, and a sense of purpose among the ranks. From the perspective of post-Second World War American military thought, these same leaders advocated new approaches to warfighting and promoted greater initiative among junior commissioned officers and senior enlisted service members. They embraced and resurrected the offensive spirit that had temporarily gone dormant.

It took at least a decade for a doctrine that integrated air–ground maneuver, the operationalization of air–ground cooperation, and information-enabled weapons systems to evolve into a new, joint approach to warfighting. The cultural sensitivity and operational outlook required to implement the envisioned dynamic, integrated, rapid-dominance style of warfare continues to mature. Doctrine, technology, and organizational innovations, moreover, retained their currency in part because they evolved in an organizational context that favored initiative and operational flexibility. The fungibility of information technology and 'how-to' knowledge about its applications was increasingly embedded in institutional practices and cognitive schemas.

Then, as now, innovative planners sought to design flexibility into future operational capabilities because of the uncertainty present in the security environment. Operational flexibility did not manifest itself until the late 1990s, which is when some planners in the 1980s actually projected that precision nonnuclear strike, intelligence capabilities, and maneuver doctrine would coalesce into a deployable Follow On Forces Attack system. The end of the Cold War delayed the actual fielding of the full range of capabilities. As was anticipated by those that emphasized capability packages over individual systems, however, the essence of the offset strategy was adapted to meet emerging post-Cold War threats after nearly a decade of inattention to modernization.

Operational flexibility is now more accepted as a modernization precept than it was during the late 1970s and 1980s. Then, flexibility and agility increased in defense policymaking, research and development, and doctrinal change; the flexibility actually envisioned in military operations paled in comparison to the agility required in the early 2000s. Now, American forces are pursuing even greater flexibility in military activities. Intratheater airlift is one critical example. Whereas the 1970s and 1980s envisioned prepared operating bases and traditional airlift requirements, operations in Afghanistan and Iraq required the rapid movement of armored brigades to austere operating bases and in-flight preparation of intelligence support. Still, it is not clear that the emerging strategic planning environment is engendering the degree of creativity and risk taking that existed among research and development organizations in the early 1980s.

Organizational factors

The security environment, perceived innovation attributes, and enabling characteristics are closely linked to organizational factors in the proposed innovation framework. In the real world, of course, these analytic distinctions fade. Perceptions of the security environment, for example, are part of a larger overlapping flow of influences that are unconstrained by what scholars label behavioral, interpersonal, or structural boundaries. Scholars reduce the complex milieu of agency and structural conditions into frameworks to facilitate analysis and render judgments. Part of the reason for the innovation milieu construct, therefore, is to focus on both the deconstruction (and reduction) of reality into manageable parts to facilitate the study of military innovation; doing also helps orient military innovation studies away from narrow frameworks that privilege only parts of the larger milieu, claiming that only one sector or set of factors is important. We want to consider the parts in the light of the whole.

Not all of the organizational factors involved in the emergence of the new American way of war can be summarized here. Students of military innovation must continue to press for understanding of the formal and informal structural constructs that cohere in the form of organizational factors. Organizational factors remain important contextual influences shaping how organizations defined and redefined missions and operating procedures in response to both international and domestic influences. And they condition innovation diffusion and adoption processes through which new ideas, technology, and operational approaches emerge, prove themselves, and displace established practices.

The military's 1970s training revolution reflected a shift in organizational priorities in the aftermath of Vietnam. Closely related were doctrinal shifts, some specifically aiming to integrate new technology and weapons systems. The Army's creation of the Training and Doctrine Command (TRADOC), for example, stemmed from recognition among senior leaders that doctrine and tactics required reinvention to push new procedures and to integrate new technology. Subsequent Army and Air Force decisions to cooperate on interdiction reflected organizational acceptance of relationships required to both meet the operational threat and facilitate development of envisioned weapons systems. The Army needed Air Force acceptance for long-range, deep strike missiles and cooperation in the targeting mission. The Air Force, on the other hand, needed Army anti-aircraft support to defeat Soviet tactical aviation and to suppress enemy air defense with long-range fires.

Both services needed research and development assistance to bring new technologies to the fight. Malcolm S. Currie's 1973 decision to reorient the Defense Advanced Research Project Agency (DARPA) reflected a national focus on pursuing technologies to resolve strategic and operational challenges. Across Stealth, Assault Breaker, and information technology projects, DARPA worked closely with the Services to understand how new doctrine and technology could be applied to operational problems in ways yielding new measures of strategic effectiveness.

Students of military innovation should also consider how, and to what degree the security environment, perceptions of innovation, and enabling resources influence organizational plans and expectations. Military innovations pulled into organizations from the outside often establish path dependencies when they condition relationships between proposed innovations and the effectiveness of the organization. Organizationally, however, it is difficult for innovation champions to achieve buy-in for new ideas or capabilities that diverge from established practices without processes for discovering organizational benefits and proving them in realistic experiments. This was certainly true for maneuver warfare doctrine, Stealth, and Assault Breaker. Each of these evolved within military organizations only after proponents successfully argued their utility for mitigating or reducing risk posed by Soviet capabilities such as the Operational Maneuver Group and a very capable, integrated air defense network. Only after the domestic political context changed to support conventional warfare innovations, moreover, would such arguments make headway.

Throughout, integration emerged as a more important theme in US military thought and defense planning. A key part of the offset strategy, integration initially concerned concepts, changes in strategic doctrine, and new conventional initiatives aiming to strengthen the relationship between nuclear and nonnuclear forces.

In addition to largely overlooking the evolution of the offset strategy, histories of advanced US conventional forces sometimes overlook the origins of arguments for an integrated deep strike, rapid dominance approach. Only recently have analysts turned to the integration of intelligence, surveillance, and reconnaissance (ISR) capabilities with organizational and doctrinal innovations.

Tightening and adapting the relationship between operations and intelligence emerged as an integration theme in the early 1980s. This included ISR capabilities developed specifically to raise the nuclear threshold in Europe by strengthening the deterrence relationship between conventional and theater nuclear weapons. By the end of the 1980s, this relationship matured such that Soviet observers viewed US conventional forces as capable of "strategic theater" operations. US military planners began using the term strategic *nonnuclear* strategic strike. This trend was reinforced in the late 1990s as operations demanded more precise geopositioning. A 1998 Defense Intelligence Agency report, for example, concluded "that precision strike weapons demand precise intelligence" and the ability "to operate effectively in the high-tempo, complex, and more lethal battlefields of the future."[16]

This study does not assess the myriad advances in satellite communications and other space-based capabilities that occurred over the past three decades. It is important, nonetheless, to note that American Military innovation relied in large part on the communications, geopositioning, surveillance, and guidance systems that exploit the coverage, perspective, timeliness, and access over denied areas gained by locating capabilities in space.

Among the benchmarks was the 1980 launch of Intelsat 5, an American communications satellite able to simultaneously relay two color television signals and some twelve thousand telephone calls. Operation Iraqi Freedom in 2003 was made possible in part because the United States leased large amounts of band-width from commercial space telecommunications providers. Integrating space with other domains of operations remains an organizational priority among all the Services and a key source of operational innovation.

Another form of integration brought together the combat arms (armor, infantry, artillery, aviation), capabilities for joint command and control, shared pursuit of common weapons systems, and mutually supportive doctrine. Such organizational issues were perhaps the most important and far-reaching of the proliferating integration thrusts that continue to be a primary axis of defense transformation in the 2000s.

The most important integration theme concerned information systems. Indeed, information technology figured prominently in successive visions for addressing strategic and operational challenges over the last three decades. The promise of information technology was increasingly linked to operational approaches to deter, defeat, and now dominate adversaries. *Washington Post* reporter Vernon Loeb observed in December 2002 that, "It took years, and increasingly impres-sive proof on the battlefield, before these inspirations were recognized for what they were—a new way of fighting that would change the calculations of war and peace in unprecedented, and still uncertain ways."[17]

Integration-focused concept demonstrations like Assault Breaker were essential for gaining insights into cumulative, some would say emergent, outcomes from combining traditional warfighting capabilities with informa-tion technology. It is not surprising that the primary architect of the offset strategy, former Secretary of Defense William J. Perry, was an engineer with experience in integrating systems when he directed defense research and engi-neering in the late 1970s. Indeed, Perry was instrumental in the development of an important signals intelligence satellite system in the 1960s that funda-mentally altered the effectiveness of US collection against a range of sensitive targets.

He was also a systems integrator who understood the travails of large project management. Systems engineering (SE) and integration (SI) capabilities are key socio-technical enablers of the shift in military effectiveness associated with the end-to-end precision strike capabilities that include intelligence, surveillance, and reconnaissance systems. They represent an important organizational approach to delivering effective military capabilities; SE/SI skills are unsung organizational enablers, reflective of a larger approach to problem solving that helped the United States win the Cold War.

Modern systems engineering capabilities evolved from the work of Brigadier General Bernard Schriever in the 1950s. He "introduced a systems approach to long-range planning that involved the analysis of potential military threats to the United States and a design for Air Force responses using advanced technology."[18] Another key figure was Simon Ramo, who "made an original contribution to the development of systems engineering by creating an organization dedicated to scheduling and coordinating the activities of a large number of contractors engaged in research and development and in testing the components and subsystems that are eventually assembled into a coherent system."[19]

Developed in the 1950s and 1960s, systems engineering and integration practices were created to manage large, Cold War projects like the SAGE (Semi-automatic Ground Environment) air defense project, the Atlas inter-continental ballistic missile, and the Polaris submarine launched missile. The Atlas Project of the 1950s was indeed a watershed, leading to a new "mode of management" that "changed the complexion of both the Cold War and the aero-space industry."[20] Where the Soviet Union failed to perfect skills needed to develop, field, and integrate complex weapons systems, "America avoided the same fate because it was more efficient... in combining complex technologies into weapon systems and integrating advanced weapons systems into its fielded forces."[21]

Information technology suggested new possibilities for operational and organizational innovation. Organizational boundaries preventing information sharing and collaboration became impediments to increasing the strategic effec-tiveness of US conventional forces. Experience in creating, evolving, and adapt-ing systems engineering and integration capabilities as a discipline facilitated the emergence of processes to help leaders identify where new technology was needed. It also positioned the United States to mature the information revolution within military organizations as new information-enabled capabilities emerged during the 1980s.

SE and SI practices emerged and were perfected in large-scale, complex systems for strategic missile defenses and nuclear command and control. SE, in particular, evolved as a distinct socio-technical approach to problem solving that more closely linked mission needs and warfighting requirements to research and development, program planning, and capability insertion than at any other time in military history. Nuclear launch warning networks and associated nuclear command and control systems evolved in response to specific needs to identify an enemy nuclear missile launch and rapidly set in motion a US retaliation. This was the heart of deterrence. Over time, the cornerstone of deterrence became not the individual weapons platforms and guidance systems but the network that linked intelligence and surveillance capabilities with nuclear weapons release processes facilitating assured retaliatory strikes.

Revisiting the innovation milieu

Widespread support for conventional modernization emerged during the mid- to late-1970s. NATO planners recognized that existing modernization efforts were insufficient to counter Soviet advances. Diplomatic initiatives proved unsuccessful in moderating what the West perceived as Soviet foreign policy adventurism coupled with increased defense spending. Economic conditions improved, especially in the United States, dampening domestic opposition to defense spending.

Underlying all of this was an important organizational development in the US Army. Paralleling the quest for renewed innovation in DARPA and other research and development arenas, and reflective of the approach to innovation pursued by Generals DePuy, Starry, and Gorman, senior Army leadership on the front lines in Europe encouraged innovation at all levels. A mix of innovation activities ensued.

Chapter 2 argued that military innovation theory is primary focused on significant innovations that alter the course of military history. Often they focus on what might be called the "big bets." Qualifying as a big bet, or true "game changing" development, requires a fundamental shift in the core competencies or missions of an organization that allows one to dominate an adversary. They often change the character of warfare.

Failure to innovate was not a primary concern of this study, although examples were mentioned. Support for unmanned aerial vehicles (UAVs) and precision munitions evolved slowly within the Air Force despite operational demonstrations of their potential. The F-117 Stealth aircraft was not accepted by the Air Force until the Chief of Staff had assurances it would not affect funding for other aircraft. Many initially balked at the Army's decision to adopt a maneuver-oriented approach along with increased reliance on air ground cooperation. Few senior military officer embraced jointness. Congressional legislation mandated by statute the reforms many defense insiders long recognized as fundamental to raising American military effectiveness.

The global positioning system (GPS) is another interesting example of how changes evolve given its revolutionary influence on military effectiveness. Initiated in 1973, GPS suffered through proposed budget cuts, survived several attempts to kill the program altogether, and achieved initial operational status only after years of delay following the Challenger space shuttle disaster. After failing to receive support from the Services, civilians in the Office of the Secretary of Defense rescued the program in the early 1980s over the objections of senior military officers. Even after initial capabilities became available in 1991, many military leaders questioned its usefulness. By the end of the decade, GPS was critical across the spectrum of military activities.

Air Launched Cruise Missile (ALCM) and Submarine Launched Cruise Missile (SLCM/Tomahawk) development remain case studies into innovation processes and the politics of weapons programs. Cruise missiles had been developed since the 1950s, competing for leadership attention and R&D

funding with ballistic missiles. Navy cruise missile programs included the Regulus I and II. Air Force programs included the Matador, Navaho, Snark, Mace, and Hound Dog. Technological and organizational factors favored ballistic missiles. Missiles were not only faster, rendering them less vulnerable to air defenses, they were more accurate and could carry a larger payload. And they were less of a threat to manned bombers, which in the 1950s remained the soul of the Air Force.

Then the strategic context changed. Arms control treaties helped shape the strategic landscape. The May 1972 Strategic Arms Limitations Talk (SALT) I agreement did not limit cruise missiles. The Soviets had them. After Soviet advances in ballistic missiles threatened US nuclear superiority, cruise missiles became an attractive option for maintaining parity. Additionally, Soviet air defense improvements threatened manned bombers. The 1973 Yom Kippur War demonstrated the capabilities of Soviet weapons systems, including air defense innovations threatening an Air Force core competency: manned strategic bombing. Low-flying cruise missiles could attack enemy air defenses. The bombers could then get through.

Existing technological developments helped. Turbofan engines evolved in the 1960s out of an Advanced Research Project Agency initiative for a jet-powered backpack. It yielded a low-cost engine used on the first mass-produced cruise missile.[22] Another key development was Terrain Contour Mapping (TERCOM). Patented in 1958, the technology enabling this navigation and guidance capability evolved through successive improvements in accuracy. TERCOM works by loading a digital map into the cruise missile guidance computer along with the intended flight path. An on-board altimeter compares the elevation of the terrain passing underneath with the digital map, computing speed and location. Corrections can be sent to the missile in-route. The guidance system "did not become feasible until advances in large-array microelectronics in the late 1960s permitted the storage of large amounts of data in small spaces with minimal power requirements."[23]

In the end, according to Henry Levine, substantive cruise missile development occurred only in periods of strategic crisis, such as the fear of a Soviet first strike advantage, during which normal weapon development routines were disrupted. This permitted extra-service organizations with interests in promoting cruise missiles and their associated technology to exercise important influence. A new "action channel" was momentarily created and exploited to achieve a reorientation of ongoing service-sponsored programs.[24]

By the 1990s, after some forty years of development, the Tomahawk emerged as a key component in the US military arsenal. It evolved as a somewhat disruptive capability in terms of challenging long-standing Air Force antipathy to unmanned strike. Unmanned Aerial Vehicles suffered from similar organizational antipathy until the early 2000s, when their utility was proven in Afghanistan and Iraq—leading to Air Force support for armed UAVs.

Given failures to innovate, it is important to clarify that merely understanding the security environment does not guarantee successful innovation. Nor does it yield understanding of specific innovation processes and outcomes. It does, however, provide a critical first step for those attempting to enact change and for those studying innovation behavior. Both involve comparing perceived threats and opportunities for meeting them with defense plans, training regimes, force structure, doctrine, and other indicators of future battlefield behavior. A point for scholars and practitioners: because this process is human, it is flawed. For this reason, any innovation framework must consider the social aspects of innovation diffusion and adoption within organizations.

A final theme from the thirty-year transformation

Motivating this study of military innovation and the origins of a new American way of war was an interest in expanding sources of historical perspective that students of US defense strategy might tap to understand the current period of defense transformation. This includes awareness of the formative and maturation phases of a new American way of war and the need to expand the range of case studies available to students of military innovation.

Military affairs span across disciplines: military history, autobiography, psychology, theory building case studies of political-military decision making, arms proliferation, political-economic studies of war making potential, international relations theories derived from correlates of war databases, action–reaction phenomena, and so on. This is a mixed blessing for scholars and decision makers looking for empirically derived insights into military innovation phenomena. As the study and practice of military change management necessarily involves understanding multifaceted contextual elements, a cross-disciplinary approach is needed. This seems particularly true when considering the strategic aspects of military change management and relationships between strategic and operational necessity and innovation activities.

In their survey of the military effectiveness of nine military organizations in the early twentieth century, Williamson Murray and Allan Millet found that "nations that got their strategy right were able to repair tactical and operational deficiencies in their military organizations. But nations that got the strategy wrong, no matter how effective their military organizations on the battlefield, *always lost*."[25] The very issue of an "innovation strategy" as part of an overall defense transformation strategy is difficult for many to comprehend.

Military history is replete with failed transformations and cases where military organizations fail to adapt. In many cases, leaders simply got the strategy wrong or failed to effectively communicate it. Other times the strategy was not implemented correctly. Increasing the overall strategic effectiveness of US national security processes will remain an elusive goal without a more comprehensive national strategy for innovation. Although pieces of this strategy exist in the form of national security plans, defense transformation visions, and homeland security reforms, no unifying study of the evolving strategic context for innovation has begun. For the still disparate arms of US national security, this leaves organizations without a clear template to prioritize innovation activities relative to current and future needs.

Defense transformation is fundamentally a strategic planning and execution process. There is no lack of strategic planning approaches suitable to the task. Designing and faithfully implementing a strategic plan that aligns resources to achieve the optimal capabilities for the situation is critical. When this requires successfully adopting disruptive or a significantly different technology or doctrine, the strategy must address diffusion and adoption processes early in the process with an eye toward agile implementation. An innovation strategy is needed.

This is where innovation studies and frameworks are useful, something businesses are discovering after a decade of re-engineering and process revolution models insensitive to end-to-end aspects of innovation. Compared with RMA-associated defense modernization policies in the early 1990s, initiatives associated with official defense transformation strategy in the 2000s reflect greater sensitivity to factors identified in strategic innovation literature as critical for successful change management. These include cultural change, strategic communication, the identification of risk associated with change, and the importance of winning the battle of perception—which includes managing expectations.

Measurement in the realm of military innovation is neither elegant nor refined, and comparative studies of different innovation cases do not easily succumb to methodological rigor or the aesthetics of metrics-based marketing research. Impacts on the environment are decidedly nonlinear, knowable only by virtue of the promise of, and potential for, ameliorating the challenges and problems inherent in the strategic or operational necessity driving the impetus to innovate. As one organizational theorist concluded, there cannot be "one best way to innovate because the innovation process is inherently probabilistic and because there are myriad forms and kinds of innovations."[26] Others argue that, "observed processes cannot be reduced to a simple sequence of stages or phases as most process models in the literature [of innovation] suggest."[27]

A single innovation theory explaining all cases remains unobtainable. As a body of work, innovation studies generally reinforce the approach taken here: the need to understand and focus on the innovation milieu within which each innovation case is nested. Moreover, they reinforce the need for an organizing theoretical framework that leaves sufficient room for incorporating disparate theoretical resources attuned to different elements of innovation existing across innovation cases and periods.[28]

John Lewis Gaddis concludes that, "studying the past has a way of introducing humility—a first stage toward gaining detachment—because it suggests the continuity of the problems we confront, and the unoriginality of most of our solutions for them. It is a good way of putting things in perspective, of stepping back to take in a wider view."[29] True enough. His perspective is important on many levels. It certainly applies to current efforts to revise American defense strategy and military thought to address small wars and counterinsurgencies.

In addition to infusing additional background information into the continuously evolving discussion of American military innovation, historical perspective on the antecedents to current technologies and operational concepts provides a sense of historical continuity into ongoing defense transformation decisions. It demonstrates how much has not changed in terms of visions for future warfighting capabilities at the same time as suggesting paths of divergence demonstrated in recent conflicts. This is where military innovation studies offer to fill an important niche in policy analysis.

Notes

1 Military innovation and defense transformation

1 Richard H. Van Atta *et al.*, *Transformation and Transition: DARPA's Role in Fostering an Emerging Revolution in Military Affairs, Volume 1—Overall Assessment* (Alexandria, VA: Institute for Defense Analyses, April 2003), p. 10.
2 Perry discussing his tenure as Undersecretary of Defense in Ashton B. Carter and William J. Perry, *Preventive Defense: A New Security Strategy for America* (Washington, DC: Brookings Institute Press, 1999), p. 180.
3 Ibid.
4 Colin S. Gray, *Strategy for Chaos: Revolutions in Military Affairs and the Evidence of History* (London: Frank Cass, 2002), p. 247.
5 William A. Owens, "Creating a U.S. Military Revolution" in Theo Farrell and Terry Terriff (eds), *The Sources of Military Change: Culture, Politics, Technology* (Boulder, CO: Lynnee Rienner, 2002), p. 207.

2 On military innovation

1 See Mathew Evangelista, *Innovation and the Arms Race: How the United States and the Soviet Union Develop New Military Technologies* (Ithaca, NY: Cornell University Press, 1988). The primary work in the field, and the one of most concern to this study, is Stephen Peter Rosen's *Winning the Next War: Innovation and the Modern Military* (Ithaca, NY: Cornell University Press, 1991).
2 See David E. Johnson, *Fast Tanks and Heavy Bombers: Innovation in the U.S. Army, 1917–1945* (Ithaca, NY: Cornell University Press, 1998); Timothy Moy, *War Machines: Transforming Technologies in the U.S. Military, 1920–1940* (College Station: Texas A & M University Press, 2001).
3 Colin S. Gray, *Strategy For Chaos: Revolutions in Military Affairs and the Evidence of History* (London: Frank Cass, 2002), p. 6.
4 Barry Watts and Williamson Murray, "Military Innovation in Peacetime" in Williamson Murray and Allan R. Millett (eds), *Military Innovation in the Interwar Period* (New York: Cambridge University Press, 1996), p. 371.
5 Williamson Murray, "Innovation Past and Future" in Murray and Millett (eds), *Military Innovation in the Interwar Period*, p. 300.
6 Allan R. Millet, "Patterns of Military Innovation in the Interwar Period" in Williamson Murray and Allan R. Millett (eds), *Military Innovation in the Interwar Period* (New York: Cambridge University Press, 1996), p. 335.
7 Murray, "Innovation: Past and Future" in Murray and Millet, p. 312.
8 Eugene Gholz, "Military Efficiency, Military Effectiveness, and Military Formats," paper presented at the 2003 annual meeting of the American Political Science Association, Philadelphia, PA, p. 2.

9 Azar Gat, *The Development of Military Thought: The Fourteenth Century* (New York: Oxford University Press, 1991), p. 247.

10 Nigel Nicholson (ed.), *The Blackwell Encyclopedic Dictionary of Organizational Behavior* (Cambridge, MA: Blackwell Business, 1995), pp. 233, 234.

11 Barry R. Posen, *The Sources of Military Doctrine: France, Britain, and Germany Between the World Wars* (New York: Cornell University Press, 1984), p. 47.

12 James Q. Wilson, *Bureaucracy: What Government Agencies Do and Why They Do It* (New York: Basic Books, Inc., 1989), p. 222.

13 Ibid., p. 225.

14 Stephen Peter Rosen, "New Ways of War: Understanding Military Innovation," *International Security* (Summer 1988), p. 134.

15 See Michael L. Tushman and Charles A. O'Reilly III, *Winning Through Innovation: A Practice Guide to Leading Organizational Change and Renewal* (Cambridge, MA: Harvard Business School Press, 2002) and Frances Hesselbein and Rob Johnston (eds), *On Creativity, Innovation, and Renewal* (San Francisco, CA: Jossey-Bass, 2002).

16 Drucker quoted in Frances Hesselbein, Marshall Goldsmith, and Iain Somerville in "Introduction" to their edited *Leading for Innovation: And Organizing for Results* (San Francisco, CA: Jossey-Bass, 2002), p. 1.

17 Michael E. Porter, *The Comparative Advantage of Nations* (New York: The Free Press, 1990), p. 49.

18 Murray and Knox, "The Future Behind Us" in MacGregor Knox and Williamson Murray (eds), *The Dynamics of Military Revolution, 1300–2050* (New York: Cambridge University Pres, 2001), p. 180.

19 Richard K. Betts, "Conventional Strategy: New Critics, Old Choices" in *International Security* (Vol. 7, No. 4), p. 149.

20 William A. Owens, "The Once and Future Revolution in Military Affairs," *Joint Forces Quarterly* (Summer 2002), p. 56.

21 Ibid.

22 Robert McC. Adams, *Paths of Fire: An Anthropologist's Inquiry into Western Technology* (Princeton, NJ: Princeton University Press, 1996), p. 5.

23 For similar comments on interactive innovation systems, albeit described with different terms, see Allan R. Millet, Williamson Murray, and Kenneth H. Watman (eds), "The Effectiveness of Military Organizations" in *Military Effectiveness, Volume I: The First World War* (Boston, MA: Unwin Hyman, 1988), p. 3; Williamson Murray, "Innovation Past and Future" in Williamson Murray and Allan R. Millett (eds), *Military Innovation in the Interwar Period* (New York: Cambridge University Press, 1996), p. 302; Allan R. Millet, "Patterns of Military Innovation in the Interwar Period" in Murray and Millett, p. 367; and Barry Watts and Williamson Murray, "Military Innovation in Peacetime" in Murray and Millett, p. 381.

24 Williamson Murray, "Innovation Past and Future," p. 305.

25 Nicholson, p. 234.

26 For an extended discussion of contextual factors in international relations theory, and a useful theoretical framework drawing on contextual and structural factors, see Gary Goertz, *Contexts of International Politics* (New York: Cambridge University Press, 1994).

27 Nathan Rosenberg, *Exploring the Black Box: Technology, Economics, and History* (Cambridge, MA: Cambridge University Press, 1994), pp. 68, 69, 70.

28 Ibid.

29 Ibid., p. 115.

30 Grant Hammond, *The Mind of War: John Boyd and American Security* (Washington, DC: Smithsonian Institution Press, 2001); James C. Burton, *The Pentagon Wars: Reformers Challenge the Old Guard* (Annapolis, Maryland: Naval Institute Press, 1993); Kenneth L. Adelman and Norman R. Augustine, *The Defense Revolution: Intelligent*

Downsizing of America's Military (Lanham, MD: Institute for Contemporary Studies, 1990).

31   Robert Buderi, *The Invention that Changed the World: How A Small Group of Radar Pioneers Won the Second World War and Launched a Technological Revolution* (New York: Simon and Schuster, 1996); Harvey M. Sapolsky, *The Polaris System Development: Bureaucratic and Programmatic Success in Government* (Cambridge, MA: Harvard University Press, 1972).

32   William Odom, *After the Trenches: The Transformation of U.S. Army Doctrine, 1918–1939* (College Station, TX: Texas A & M University Press, 1999), pp. 244–245.

33   James Kitfield's *Prodigal Soldiers* (New York: Simon and Schuster, 1995).

34   Richard K Betts (ed.), *Cruise Missiles: Technology, Strategy, Politics* (Washington, DC: The Brookings Institution, 1981).

35   Allan R. Millet and Williamson Murray, *Military Effectiveness*, 3 Vols (Allen and Unwin, 1988).

36   MacGregor Knox and Williamson Murray's edited volume, *The Dynamics of Military Revolution, 1300–2050* (New York: Cambridge University Press, 2001).

37   Chapter 8 in Knox and Murray, *The Dynamics of Military Revolution, 1300–2050*.

38   Benjamin S. Lambeth, *The Transformation of American Air Power* (Ithaca, NY: Cornell University Press, 2000).

39   Richard P. Hallion, *Storm Over Iraq: Air Power and the Gulf War* (Washington, DC: Smithsonian Institute Press, 1992).

40   Kenneth P. Werrell, *Chasing the Silver Bullet: U.S. Air Force Weapons Development from Vietnam to Desert Storm* (Washington, DC: Smithsonian Books, 2003).

41   Michael Russell Rip and James M. Hasik, *The Precision Revolution: GPS and the Future of Aerial Warfare* (Annapolis, MS: Naval Institute Press, 2002).

42   Snyder, *The Ideology of the Offensive: Military Decision Making and the Disasters of 1914* (Ithaca, NY: Cornell University Press, 1984); Posen (Ithaca, NY: Cornell University Press, 1984); Elizabeth Kier, *Imaging War: French and British Military Doctrine Between the Wars* (Princeton, NJ: Princeton University Press, 1997).

43   Posen, p. 13.

44   Ibid., pp. 141–178.

45   Kier, p. 20.

46   Ibid., p. 23.

47   Williamson Murray and Allan R. Millett (eds), *Military Innovation in the Interwar Period* (Cambridge, MA: Cambridge University Press, 1996). See especially Millett "Patterns of Military Innovation in the Interwar Period," pp. 329–368. Farrell, Theo, and Terry Terriff (eds), *The Sources of Military Change: Culture, Politics, Technology* (Boulder, CO: Lynn Rienner Publisher, 2002).

48   Katzenbach in Endicott, John E., and Stafford, Roy W (eds), *American Defense Policy*, fourth edition (Baltimore: Johns Hopkins University Press, 1977), pp. 360–373.

49   Armacost, Michael H., *The Politics of Weapons Innovation: The Thor-Jupiter Controversy* (New York: Columbia University Press, 1969).

50   Stephen Peter Rosen, *Winning the Next War: Innovation and the Modern Military* (Ithaca, NY: Cornell University Press, 1991).

51   Ibid., p. 1.

52   Ibid., p. 251.

53   Ibid., p. 52.

54   Ibid., pp. 39–40.

55   Ibid., p. 110.

56   Ibid., p. 253.

57   The business world's focused on 'revolutionary change' approaches to management in the early 1990s waned at the end of the decade as manegement gurus and business process re-engineering studies realized that revolutionary approaches tendes to fail. Overall, as the decade progressed, the "hard right turn" philosophy waned in favor of

transformation and innovation, with management strategies becoming more attuned to the diffusion of innovations through an organization's culture and processes. Among the most important changes was greater attention to culture, strategic communications, and understanding the relationship between leadership and followership. For a summary of this, see "The HBR List: Breakthrough Ideas for Today's Business Agenda," *Harvard Business Review* (April 2001), p. 125.

58 Ibid.
59 Thomas Kuczmarski, Arthur Middlebrooks, and Jeffrey Swaddling, *Innovating the Corporation: Creating Value for Customers and Shareholders* (Chicago, IL: NTC Business Books, 2001), pp. 20–21.
60 John D. Wolpert, "Breaking out of the Innovation Box," *Harvard Business Review* (August 2002), p. 77.
61 James Champy, *X-Engineering the Corporation: Reinventing Your Business in the Digital Age* (New York: Warner Books, 2001), pp. 2–3.
62 Paul C. Light, *Sustaining Innovation: Creating Nonprofit and Government Organizations the Innovate Naturally* (San Francisco, CA: Jossey-Bass Publishers, 1998), p. xiv.
63 John Kao, "Reinventing Innovation" in Hesselbein, Goldsmith, and Somerville, p. 275.
64 Ibid., p. 275.
65 Clayton M. Christensen, *The Innovator's Dilemma: When New Technologies Cause Great Firms to Fail* (Boston: Harvard Business School Press, 1997), p. xiii.
66 Ibid., p. xv.
67 Peter F. Drucker, "Really Reinventing Government" in the *Atlantic Monthly* (February 1995), p. 50.
68 Kao, p. 284.
69 Michael Roberts, *The Military Revolution, 1560–1660* (Belfast: Queen's University Press, 1956). The most comprehensive review of the literature is Clifford Rogers (ed.), *The Military Revolution Debate* (Boulder, CO: Westview Press, 1995).
70 Summarized by Alex Roland in "Technology and War: The Historiographical Revolution of the nineteen-eighties," *Technology and Culture* (Vol. 34, No. 1), p. 123.
71 Geoffrey Parker, *The Military Revolution: Military Innovation and the Rise of the West, 1500–1800* (New York: Cambridge University Press, 1988).
72 Brian M. Downing, *The Military Revolution and Political Change: Origins of Democracy and Autocracy in Early Modern Europe* (Princeton, NJ: Princeton University Press, 1992).
73 Ibid., p. 10.
74 *The Journal of Military History* (Vol. 57, No. 2), p. 242.
75 Ibid., p. 276.
76 Williamson Murray, "Introduction," in Williamson Murray (ed.), *The Emerging Strategic Environment* (Westport, CT: Praeger, 1999), p. xxxiv.
77 Watts and Murray, "Military Innovation in Peacetime" in Williamson Murray and Allan R. Millet (eds), *Military Innovation in the Interwar Period* (New York: Cambridge University Press, 1996), p. 405.
78 Williamson Murray, "May 1940: Contingency and Fragility of the German RMA," MacGregor Knox and Williamson Murray (eds), *The Dynamics of Military Revolution, 1300–2050* (New York: Cambridge University Press, 2001), p. 155.
79 William J. Perry, "Military Action: When to Use and How to Ensure its Effectiveness" in James E. Nolan (ed.), *Global Engagement: Cooperation and Security in the 21st Century* (Washington, DC: Brookings Institution Press, 1994), p. 240. Also cited in William A. Owens, "Creating a U.S. Military Revolution" in Theo Farrell and Terry Terriff (eds), *The Sources of Military Change* (Boulder, CO: Lynne Reinner Publishers, 2002), p. 219.
80 Barry Watts and Williamson Murray, "Military Innovation in Peacetime" in Williamson Murray and Allan R. Millet (ed.), *Military Innovation in the Interwar Period* (New York: Cambridge University Press, 1996), p. 373.

81 Shimon Naveh, *In Pursuit of Military Excellence: The Evolution of Operational Theory* (London: Frank Cass, 1997), p. 126.
82 Ibid., p. 105.
83 Ibid., p. 128.
84 Matthew Cooper, *The German Army, 1933–1945: Its Political and Military Failure* (Lanham, MD: Scarborough House, 1978).
85 Shimon Naveh, *In Pursuit of Military Excellence: The Evolution of Operational Theory* (London: Frank Cass, 1997).

3 American military strategy from the Second Word War through Vietnam

1 Noteworthy attempts to review the Cold War or important political-military developments during it include: Ronald E. Powaski, *The Cold War: The United States and the Soviet Union, 1917–1991* (New York: Oxford University Press, 1998); Lawrence Freedman, *The Evolution of Nuclear Strategy*, second edition (New York: St. Martin's Press, 1989); John Newhouse, *War and Peace in the Nuclear Age* (New York: Alfred A. Knopf, 1989); Martin Walker, *The Cold War: A History* (New York: Henry Holt and Company, 1993); Richard Crockatt, *The Fifty Years War: The United States and the Soviet Union in World Politics, 1941–1991* (New York: Routledge, 1995); and David Reynolds, *One World Divisible: A Global History Since 1945* (New York: W.W. Norton and Company, 2000). Informative accounts of early U.S. defense planning include: Thomas D. Boettcher, *First Class: The Making of the Modern U.S. Military, 1945–1953* (Boston: Little, Brown, and Company, 1992); Amy B. Zegart, *Flawed By Design, The Evolution of the CIA, JCS, and NSC* (Stanford, CA: Stanford University Press, 1991); Warner R. Schilling, Paul Y. Hammon, and Glenn H. Snyder, *Strategy, Politics, and Defense Budgets* (New York: Columbia University Press, 1962); and, Maurice A. Mallin, *Tanks, Fighters, and Ships: U.S. Conventional Force Planning Since WWII* (New York: Brassey's Inc., 1990). See also the seven volume *History of the Joint Chiefs of Staff* (Washington, DC: Office of Joint History, Office of the Chairman of the Joint Chiefs of Staff). Further studies are cited later.
2 James F. Brynes, *Speaking Frankly* (New York: Harper and Brothers Publishers, 1947), p. 256.
3 Ibid.
4 Armed Forces strength information, derived from Army and Navy annual reports, is adapted from a table in James F. Schnabel, *History of the Joint Chiefs of Staff, The Joint Chiefs of Staff and National Policy, Volume I: 1945–1947* (Washington, DC: Office of the Chairman of the Joint Chiefs of Staff, 1996), p. 109 and 225, fn. 56.
5 Kennan's telegram is reprinted, with commentaries, with similar documents from British and Soviet analysts in Kenneth M. Jensen (ed.), *Origins of the Cold War: The Novikov, Kennan, and Roberts "Long Telegrams" of 1946* (Washington, DC: U.S. Institute of Peace, 2000).
6 Kennan discusses the period in George Kennan, *Memoirs, 1925–1950* (New York: Pantheon Books, 1967), chapter 11. The so-called "Long Telegram" is reprinted and discussed in Thomas H. Etzold and John Lewis Gaddis (eds), *Containment: Documents on American Policy and Strategy, 1945–1950* (New York: Columbia University Press, 1978), pp. 49–64. For information on Kennan, his telegram, and its impact on policy, see: John Lewis Gaddis, *The United States and the Origins of the Cold War, 1941–1947* (New York: Columbia University Press, 1972) and David Mayers, *George Kennan and the Dilemmas of U.S. Foreign Policy* (New York: Oxford University Press, 1988), chapters 5 and 6.
7 Analysis of Kennan's telegram and its impact is provided by John Lewis Gaddis, *Strategies of Containment: A Critical Appraisal of Postwar American National Security Policy* (New York: Oxford University Press, 1982), especially chapters 1 and 2.

8 Henry A. Kissinger, *Nuclear Weapons and Foreign Policy* (New York: Harper and Brothers, 1957), pp. 12–13.

9 John Lewis Gaddis, *We Now Know: Rethinking Cold War History* (New York: Oxford University Press, 1997), p. 100. Emphasis in original.

10 Russell F. Weigley, *The American Way of War: A History of the United States Military Strategy and Policy* (New York: Macmillan Publishing Co., Inc., 1973), part five.

11 James F. Schnabel, *History of the Joint Chiefs of Staff, The Joint Chiefs of Staff and National Policy, Volume I: 1945–1947* (Washington, DC: Office of the Chairman of the Joint Chiefs of Staff, 1996), p. 135.

12 David Alan Rosenberg, "The Origins of Overkill: Nuclear Weapons and American Strategy, 1945–1960" *International Security* (Vol. 7, No. 4), p. 38.

13 Rosenberg, p. 15.

14 For a history of the origins of atomic targeting and nuclear doctrine, see David Alan Rosenberg, "The Origins of Overkill: Nuclear Weapons and American Strategy, 1945–1960" *International Security* (Vol. 7, No. 4), pp. 3–71; Rosenberg, p. 12.

15 For more on military planning and the military services see Kenneth W. Condit, *The Joint Chiefs of Staff and National Policy: Volume II, 1947–1949* (Washington, DC: Office of the Chairman of the Joint Chiefs of Staff, 1996).

16 Office of Reports and Estimates, ORE 25–48, "The Break-Up of the Colonial Empires and its Implications for US Security" (September 3, 1948) in Michael Warner (ed.), *The CIA Under Harry Truman* (Washington, DC: Center for the Study of Intelligence, Central Intelligence Agency, 1994), pp. 223–224.

17 Melvyn P. Leffler, *A Preponderance of Power: National Security, the Truman Administration, and the Cold War* (Stanford, CA: Stanford University Press, 1992) p. 163.

18 Joint Ad Hoc Committee, "Possibility of Direct Soviet Military Action During 1948," (Central Intelligence Agency: Office of Reports and Estimates, March 30, 1948), p. 1.

19 Rosenberg, p. 13.

20 Ibid., p. 13.

21 Ibid., p. 14.

22 Report of a Joint Ad Hoc Committee, Office of Reports and Estimates, ORE 32–50, "The Effect of the Soviet Possession of Atomic Bombs on the Security of the United States" (June 9, 1950) in Warner, p. 330.

23 Rosenberg, p. 16.

24 Freedman, *The Evolution of Nuclear Strategy*, p. 71.

25 Rosenberg, p. 17.

26 Freedman, *The Evolution of Nuclear Strategy*, p. 78. See also "Appraising U.S. National Security Policy" by Daniel J. Kaufman, Jeffrey S. McKitrick, and Thomas J. Leney (eds) *U.S. National Security: A Framework for Analysis* (Lexington, MA: DC Heath and Company, 1985), p. 554.

27 NSC 162/2, October 30, 1953, p. 22.

28 Mark Trachtenberg, *A Constructed Peace: The Making of the European Settlement, 1945–1963* (Princeton, NJ: Princeton University Press, 1999), p. 158.

29 Trachtenberg, p. 163.

30 Ibid., p. 163.

31 Ibid., p. 162.

32 Ibid., p. 163.

33 "NIE [National Intelligence Estimate] 11-6-54: Soviet Capabilities and Probable Programs in the Guided Missile Field" cited by Ernest R. May, "Strategic Intelligence and U.S. Security: The Contributions of CORONA" in Dwayne A. Day, John M. Logsdon, and Brian Letein (eds), *Eye in the Sky: The Story of the CORONA Spy Satellites* (Washington, DC: Smithsonian Institution Press, 1998), p. 22.

34 For an overview of theoretical and policy considerations concerning surprise attacks, see Richard K. Betts, *Surprise Attack: Lessons for Defense Planning* (Washington, DC, 1982).

35 John Foster Dulles, "Challenges and Response in U.S. Policy," *Foreign Affairs* (Vol. 25, No. 1), p. 31.

36 Gaddis, *Origins of Containment*, p. 178.

37 Rosenberg, p. 66.

38 Desmond Ball in Ball and Richelson, p. 61.

39 Lawrence Freedman, p. 87.

40 Ambrose makes a similar argument, op. cit. p. 316.

41 Gavin, *War and Peace in the Space Age* (New York: Harper and Brothers, 1958).

42 Desmond Ball, *The Development of the SIOP, 1960–1983* in Desmond Ball and Jeffrey Richelson (eds), *Strategic Nuclear Targeting* (Ithaca, NY: Cornell University Press, 1986), pp. 59–63.

43 John Newhouse, *War and Peace in the Nuclear Age* (New York: Alfred A. Knopf, 1989), p. 162.

44 Newhouse, p. 163.

45 Mark Trachtenberg, *A Constructed Peace: The Making of the European Settlement, 1945–1963* (Princeton, NJ: Princeton University Press, 1999), p. 352.

46 Robert S. McNamara, speech before the American Bar Foundation, Chicago, Illinois, February 17, 1962 reprinted in the March 1, 1962 issue of *Vital Speeches* (Vol. 28, No. 10), pp. 296–299.

47 Samuel P. Huntington, "Conventional Deterrence and Conventional Retaliation in Europe" in Steven E. Miller (ed.), *Conventional Forces and American Defense Policy* (Princeton, NJ: Princeton University Press, 1986), p. 259; table adapted from p. 260.

48 Ball, p. 63.

49 Training manual cited by Williamson Murray, "Air Power Since World War II: Consistent With Doctrine?" in Richard H. Schultz, Jr and Robert L. Pfaltzgraff Jr (eds), *The Future of Air Power in the Aftermath of the Gulf War* (Maxwell Air Force Base, Alabama: Air University Press, 1992), p. 104.

50 Draft Presidential Memorandum (DPM) on Strategic Offensive and Defensive Forces, January 9, 1969, p. 6. Discussed in Zisk, p. 85.

51 US Congress, Office of Technology Assessment, *New Technology for NATO: Implementing Follow-On Force Attack*, OTA-ISC-309 (Washington, DC: U.S. Government Printing Office, June 1987), p. 56.

52 Joseph D. Douglass, Jr, *The Soviet Theater Nuclear Offensive* (Washington, DC: Government Printing Office, 1976), p. 15.

53 William R. Kintner and Harriet Fast Scott (eds), *The Nuclear Revolution in Soviet Military Affairs* (Norma, OK: University of Oklahoma Press, 1968), p. 400.

54 Oleg Penkovsky, *Claws of the Bear: The History of the Red Army from the Revolution to the Present* (Boston, MA: Houghton Mifflin, 1989), fn. 10, pp. 252, 444.

55 Oleg Penkovsky, *Claws of the Bear*, p. 253.

56 Kimberly Martin Zisk, *Engaging the Enemy: Organizational Theory and Soviet Military Innovation, 1955–1991* (Princeton, NJ: Princeton University Press, 1993), p. 63.

57 Robert A. Doughty, "The Cold War and the Nuclear Era: Adjusting Warfare to Weapons of Mass Destruction," Robert A. Doughty and Ira D. Gruber (eds), *Warfare in the Western World*, Vol. II, p. 858.

58 Richard Simpkin, *Race to the Swift: Thoughts of Twenty-First Century Warfare* (New York: Brassey's Defence Publishers, 1985), pp. 44–46.

59 Michael MccGwire, *Military Objectives in Soviet Foreign Policy* (Washington, DC, 1987), p. 3.

60 MccGwire, p. 29.

61 Dale R. Herspring, *The Soviet High Command, 1967–1989: Personalities and Politics* (Princeton, NJ: Princeton University Press, 1990), p. 64.

62 Joseph D. Douglass, Jr and Amoretta M. Hoeber, *Conventional War and Escalation: The Soviet View* (New York: Crane, Russak, and Company, 1981), p. 10.

63 Douglass, p. 113.

64 Douglass, p. 16, quoting A. A. Sidorenko's 1972 book, *The Offensive (A Soviet View)*, a translated edition of which is published by the US Air Force (Washington, DC: Government Printing Office, 1974).

65 Douglass, p. 17, quoting I. Zavyakov, "New Weapons and the Art of War," in *Red Star* (October 30, 1970) as translated by the Foreign Broadcasting Information Report (FBIS), *Daily Report: Soviet Union* (November 4, 1970), p. 2.

66 Steven L. Canby, report R-1088-ARPA, *NATO Military Policy: Obtaining Conventional Comparability With the Warsaw Pact* (Santa Monica, CA: The RAND Corporation, June 1973), p. 22.

67 Unclassified Directorate of Intelligence memo in Haines and Leggett, p. 11.

68 Ibid.

69 Peter G. Tsouras, *Changing Orders: The Evolution of the World's Armies, 1945 to the Present* (London: Arms and Armour Press, 1994), pp. 172–173.

70 Unclassified Directorate of Intelligence memo in Haines and Leggett, p. 11.

71 Soviet Lieutenant General G. I. Demidkov cited in Andrei A. Kokoshin, *Soviet Strategic Thought, 1917–91* (Cambridge, MA: The MIP Press, 1998), p. 178.

72 Kimberly Martin Zisk, *Engaging the Enemy: Organizational Theory and Soviet Military Innovation, 1955–1991* (Princeton, NJ: Princeton University Press, 1993), p. 48.

73 Zisk, *Engaging the Enemy*, p. 57.

74 McGovern cited in Marc Fisher, "Reopening the Wounds of Vietnam," *Washington Post National Weekly Edition* (March 1–7, 1995), p. 10.

75 Colby remarks at a Congressional panel on US security policies, "Documentation: U.S. National Security," 1977–2001 in *International Security* (Vol. 2, No. 2), p. 171 [171–183].

76 Hughes interview, March 7, 2003.

77 Martin Van Creveld, *Nuclear Proliferation and the Future of Conflict*, pp. 53–54.

78 Ibid., p. 59.

79 Freedman Second Edition (New York: St. Martin's Press, 1989), p. 433.

80 Bernard Brodie (ed.), *The Absolute Weapon: Atomic Power and World Order* (New York: Harcourt, Brace, 1946).

81 Ibid., p. 76.

4 Military innovation in the shadow of Vietnam: the offset strategy

1 William Perry, "Desert Storm and Deterrence," *Foreign Affairs* (Vol. 70, No. 4), p. 68.

2 Kimberly Martin Zisk, *Engaging the Enemy: Organizational Theory and Soviet Military Innovation, 1955–1991* (Princeton, NJ: Princeton University Press, 1993), pp. 75–76.

3 William J. Perry, "Desert Strom and Deterrence," *Foreign Affairs* (Vol. 70, No. 4), p. 81.

4 Ibid., p. 69.

5 Alain C. Enthoven, "U.S. Forces in Europe: How Many? Doing What," p. 514.

6 Address by Sam Nunn to the New York Militia Association September 11, 1976 reproduced in the "Documentation" section, *Survival* (Vol. 19, No. 1), p. 30.

7 Henry Kissinger, *The White House Years* (Boston, MA: Little, Brown, and Company, 1979), p. 215.

8 John Lewis Gaddis, *Strategies of Containment: A Critical Appraisal of Postwar American National Security Policy* (New York: Oxford University Press, 1982), p. 322.

9 Gaddis, *Strategies of Containment*, p. 322.

10 Ibid., p. 323.

11 Kissinger, *White House Years*, p. 33.

12 Gaddis, *Strategies of Containment*, p. 323.

13 Ibid.

14 Thomas B. Cochran *et al.*, *Soviet Nuclear Weapons* (New York: Harper & Row Publishers, 1989), p. 194.

15 Richard M. Swain, "AirLand Battle" in George F. Hofmann and Donn A. Starry (eds), *Camp Colt to Desert Storm: The History of U.S. Armored Forces* (Lexington, KN: University of Kentucky Press, 1999), p. 366.
16 General Alexander Haig's address to the US Army Association October 13, 1976 reproduced in "Documentation" section, *Survival* (Vol. 19, No. 1), p. 34.
17 Rowen testimony cited in *Aviation Week* (September 22, 1975), p. 51.
18 Jacob W. Kipp, "Conventional force modernization and the asymmetries of military doctrine: Historical Reflections on Air/Land Battle and the Operational Maneuver Group" in Carl G. Jacobsen (ed.), *The Uncertain Course: New Weapons, Strategies, and Mind-sets* (New York: Oxford University Press, 1987), p. 150.
19 Richard H. Van Atta, Jack H. Nunn, and Alethia Cook "Assault Breaker" in Richard Van Atta *et al.*, *Transformation and Transition: DARPA's Role in Fostering and Emerging Revolution in Military Affairs, Volume II—Detailed Assessments* (Alexandria, VA: Institute for Defense Analyses, IDA Paper P-3698, November 2003), pp. IV–9.
20 Alex Roland, "Technology and War" in *American Diplomacy* (Vol. II, No. 2), posted at www.unc.edu/depts/diplomat/AD_issues/4amdipl.html
21 Westmoreland cited in M.W. Hoag, *New Weaponry and Defending Europe: Some General Considerations* (Santa Monica, CA: The Rand Corporation, October 1973), p. 13.
22 Westmoreland articulated this vision in a 1969 speech to the Association of the US Army, reprinted in Appendix A of Paul Dickson's *The Electronic Battlefield* (Bloomington, IN: Indiana University Press, 1976), pp. 215–223.
23 Dickson, *The Electronic Battlefield*, pp. 116–117.
24 Ibid., p. 101.
25 Paul Dickson, *Sputnik: The Shock of the Century* (New York: Walker and Company, 2001), p. 194.
26 Van Atta, Deitchman, and Reed, pp. II–2.
27 Ibid., pp. II–17.
28 Interview, Richard H. Van Atta, September 5, 2003.
29 Richard H. Van Atta, Seymour J. Deitchman, and Sidney G. Reed, *DARPA Technical Accomplishments, Volume III* (Alexandria, VA: Institute for Defense Analyses, 1991), pp. II–14.
30 Ibid.
31 DARPA program manager Robert Moore cited in Richard H. Van Atta, Sidney G. Reed, and Seymour J. Deitchman, *DARPA Technical Accomplishments Vol. II: An Historical Review of Selected DARPA Projects* (Alexandria, VA: Institute for Defense Analyses, 1991), pp. 8–14.
32 General William E. DePuy correspondence to Gen. Creighton W. Abrams, January 14, 1974, in Col. Richard M. Swain (ed.), *Selected Papers of General William E. DePuy, USA Retired* (Fort Leavenworth, KS: Combat Studies Institute, 1994), p. 71. Additional background on TRADOC and the DePuy cite is also cited in Swain.
33 Van Atta interview with Moore in Van Atta *et al.* "Assault Breaker" chapter, pp. IV–6.
34 Interview, Richard Van Atta, Alexandria, VA, September 5, 2003.
35 Richard H. Van Atta, Sidney G. Reed, and Seymour J. Deitchman, *DARPA Technical Accomplishments Vol. II*, pp. 5–1.
36 Richard Van Atta, Jack Nunn, and Alethia Cook, "Assault Breaker" in Van Atta, Deitchman, and Reed, pp. IV–3. Hereafter Van Atta *et al.*, Assault Breaker chapter.
37 Van Atta *et al.*, "Assault Breaker" chapter, pp. IV–7.
38 Perry discussing his tenure as Undersecretary of Defense in Ashton B. Carter and William J. Perry, *Preventive Defense: A New Security Strategy for America* (Washington, DC: Brookings Institute Press, 1999), p. 180.
39 Perry cited in Van Atta *et al.* "Assault Breaker" chapter.
40 Ibid.
41 Van Atta interview with Fossum in Van Atta *et al.* "Assault Breaker" p. IV–10.

42 Richard Van Atta, personal communication with author, March 2004.

43 Ibid., pp. IV–14.

44 Ibid., pp. IV–18.

45 Van Atta *et al*. "Assault Breaker" chapter (2003), pp. IV–1.

46 Ibid.

47 Richard H. Van Atta, Sidney G. Reed, and Seymour J. Deitchman, *DARPA Technical Accomplishments Vol. II*, pp. 5–17.

48 Van Atta *et al*. "Assault Breaker," pp. IV–9; IV–41.

49 Richard H. Van Atta, Sidney G. Reed, and Seymour J. Deitchman, *DARPA Technical Accomplishments Vol. II*, pp. 5–1.

50 Paul H. Nitze, "Is it Time to Junk Our Nukes?" *Washington Post* (January 16, 1994), p. C1.

51 Van Atta and Michael J. Lippitz, *DARPA's Role in Fostering and Emerging Revolution in Military Affairs* (Alexandria, VA: Institute for Defense Analyses, November 19, 2001), p. 7.

52 Bill Owens, *Lifting the Fog of War* (New York: Farrar, Straus, and Giroux, 2000), p. 81.

53 Donn A. Starry, "Reflections" in Hoffman and Starry, p. 546.

54 Ibid., p. 547.

55 Alvin Paul Drischler, "General-Purpose Forces in the Nixon Budgets," *Survival* (Vol. 15, No. 3), p. 120 [119–123].

56 Edward C. Meyer, R. Manning Ancell, and Jane Mahaffey, *Who Will Lead? Senior Leadership in the United States Army* (Westport, CN: Praeger Publishers, 1995), p. 147.

57 GEN Alexander Haig's address to the US Army Association October 13, 1976 reproduced in "Documentation" section, *Survival* (Vol. 19, No. 1), p. 34.

58 Rick Atkinson, *The Long Gray Line: The American Journey West Point's Class of 1966* (Boston, MA: Houghton Mifflin, 1989), p. 366.

59 Swain in Hoffman and Starry, p. 367.

60 Carafano, p. 6.

61 The influence of these studies on Starry's thinking are reported in Martin J. D'Amato, "Vigilant Warrior: General Donn A. Starry's AirLand Battle and How It Changed the Army," *Armor* (Vol. 59, No. 3), p. 22.

62 Swain in Hoffman and Starry, p. 379.

63 Ibid., p. 378.

64 Ibid., p. 372.

65 Ibid., p. 377.

66 Meyer's correspondence with Starry quoted by Swain in Hoffman and Starry, p. 380.

67 Air Force Manual 1-1, *United States Air Force Basic Doctrine* (Washington, DC: Department of the Air Force, September 28, 1971).

68 *United States Air Force Basic Doctrine*, p. 2.

69 *United States Air Force Basic Doctrine*, chapter 6.

70 Charles J. Goss, *American Military Aviation: The Indispensable Arm* (College Station, TX: Texas A&M University Press, 2002), p. 223.

71 Gross, p. 223.

72 Robert J. Hamilton, a B-52 instructor pilot and flight commander, discusses the advent of conventional training in *Green and Blue in the Wild Blue* (Maxwell Air Force Base, Alabama: School of Advanced Air Power Studies, 1993), p. 32.

73 Personal correspondence, February 2004.

74 Richard Van Atta of the Institute for Defense Analyses and the Honorable Jacques Gansler both made this point in interviews and private correspondence.

75 Ibid., p. 10.

76 Ibid.

77 Richard Van Atta and Michael J. Lippitz, *DARPA's Role in Fostering an Emerging Revolution in Military Affairs* (Alexandria, VA: Institute for Defense Analyses, November 2001), p. 9.

78 For more information on these attacks, see: George and Meredith Friedman, *The Future of War: Power, Technology, and American World Dominance in the 21st Century* (New York: Crown Publishers, 1996), pp. 237–240; Benjamin S. Lambeth, *The Transformation of American Air Power* (Ithaca, NY: Cornell University Press, 2000), pp. 39–40; and Michael Russell Rip and James M. Hasik, *The Precision Revolution: GPS and the Future of Aerial Warfare* (Annapolis, MD: Naval Institute Press, 2002).
79 George and Meredith Friedman, *The Future of War*, p. 240.
80 Richard P. Hallion, *Storm Over Iraq: Air Power and the Gulf War* (Washington, DC: Smithsonian Institution Press, 1992), p. 305.
81 Hallion, *Storm Over Iraq*, p. 305.
82 Van Atta, Nunn, and Cook "Assault Breaker" in Richard Van Atta *et al.*, pp. IV–5.
83 Alain C. Enthoven, "U.S. Forces in Europe: How Many? Doing What" in *Foreign Affairs* (Vol. 53, No. 3), p. 513.
84 Steven L. Canby, *NATO Military Policy: Obtaining Conventional Comparability with the Warsaw Pact* (Santa Monica, CA: RAND, 1973), p. 69.
85 James Kitfield, *Prodigal Soldiers: How A Generation of Officers Born of Vietnam Revolutionized the American Style of War* (New York: Simon and Schuster, 1995), p. 175
86 Benjamin S. Lambeth, *The Transformation of American Air Power* (Ithaca, NY: Cornell University Press, 2000), p. 56.
87 Thomas Karas, *The New High Ground: Strategies and Weapons of Space-Age War* (New York: Simon and Schuster, 1983), p. 90.
88 Kenneth Allard, *Command, Control and the Common Defense* (Washington, DC: 1990), p. 147.
89 Interview, June 24, 2003.
90 Ibid.
91 Richard H. Van Atta, Sidney G. Reed, and Seymour J. Deitchman, *DARPA Technical Accomplishments Vol. II*, pp. 7–1.
92 Ibid.
93 Van Atta *et al.* "Assault Breaker," pp. IV–11.
94 Ibid.
95 Edward Waltz and James Llinas, *Multisensor Data Fusion* (Boston, MA: Artech House, 1990), p. 11.
96 Kenneth Allard, *Command, Control and the Common Defense* (Washington, DC: 1990), p. 147.
97 Ibid., p. 148.
98 See the summary of the 1973 NATO defense planning "issues" in *NATO Review* (April 1973), pp. 20–22.
99 Allard, p. 145.
100 Keegan, cited in Weiner, p. 21.
101 Albert Wohlstetter, "Threats and Promises of Peace: Europe and American in the New Era," *Orbis* (Winter 1974), p. 1124.
102 David Frum, *How We Got Here: The 1970s* (New York: Basic Books, 2000).
103 Ibid., p. 5.
104 Ibid., p. 344.
105 Gaddis Smith, *Morality, Reason, and Power: American Diplomacy in the Carter Years* (New York: Hill and Wang, 1986), p. 5.
106 See Raymond L. Garthoff, *Détente and Confrontation: American–Soviet Relations From Nixon to Reagan* (Washington, DC: Brookings Institution, 1985), pp. 563–716.
107 Carter Cited in Jacquelyn K. Davis and Robert L. Pfaltzgraff, Jr, *Soviet Theater Strategy: Implications for NATO*, USSI Report 78–1 (Washington, DC: United States Strategic Institute, 1978), p. 54.
108 Smith, *Morality, Reason, and Power*, p. 10.
109 Brzezinski cited in Christopher Coker, *U.S. Military Power in the 1980s* (London: The Macmillan Press, 1983), p. 31.

110 Harold Brown in ibid., pp. 32–33.
111 M. W. Hoag, *New Weaponry and Defending Europe: Some General Considerations* (Santa Monica, CA: RAND, October 1973), p. 14.
112 Hoag, p. 13.
113 J. F. Digby and G. K Smith, *Background on PGMs for NATO: Summarizing our Quick Look* (Santa Monica, CA: RAND, December 1973), p. 2.
114 Owens, *Lifting the Fog of War*, p. 89.
115 Ashton B. Carter, "Keeping America's Military Edge" in *Foreign Affairs* (January/February 2001), p. 99.
116 Brown cited in Lucille Horgan's doctoral dissertation, *Innovation in the Department of Defense, 1970 to 1987* (Carnegie-Mellon University, 1990), p. 152.
117 Leebaert, p. 454.
118 Keegan cited in Milton G. Wiener, *Surprise Attack: The Delicate Balance of Error* (Santa Monica, CA: RAND, 1977), p. 21.
119 Rumsfeld cited in John J. Mearsheimer, "Why the Soviets Can't Win Quickly in Central Europe" in Steven E. Miller (ed.), *Conventional Forces and American Defense Policy* (Princeton, NJ: Princeton University Press, 1986), p. 154.

5 Expanding missions, new operational capabilities

1 Interview with LTG James C. King, National Imagery and Mapping Agency, August 12, 2003.
2 Barry R. Posen, *The Sources of Military Doctrine: France, Britain, and Germany Between the World Wars* (Ithaca, NY: Cornell University Press, 1984), p. 13.
3 Christopher Pain, "On the Beach: The Rapid Deployment Force and the Nuclear Arms Race" in *MERIP Reports* (January 1983), p. 4 citing Reagan from the April 15, 1980 *The New Republic*.
4 Anderson cited in Lou Cannon, *President Reagan: The Role of a Lifetime* (New York: Public Affairs, 2000), p. 254.
5 Benjamin S. Lambeth, *The Transformation of American Air Power* (Ithaca, NY: Cornell University Press, 2000), p. 58.
6 Ralph Sanders, "Introduction" in Franklin D. Margiatta and Ralph Sanders (eds), *Technology, Strategy, and National Security* (Washington, DC: National Defense University Press, 1985), pp. 4–5.
7 Ibid.
8 Interview, Robert Schulenberg, July 14, 2003.
9 Thomas C. Reed, *At the Abyss: An Insider's History of the Cold War* (New York: Ballantine Books, 2004), p. 236.
10 Reed, p. 235.
11 Derek Leebaert, *The Fifty-Year Wound: The True Price of America's Cold War Victory* (Boston, MA: Little, Brown and Company, 2002), p. 507.
12 Cannon, 273.
13 Reed, pp. 236–237.
14 Ibid., p. 240.
15 Ronald E. Powaski, *The Cold War: The United States and the Soviet Union, 1917–1991* (New York: Oxford University Press, 1998), p. 233.
16 Reed, p. 240.
17 Derek Leebaert, p. 507.
18 Reed, p. 240.
19 Casper Weinberger, *Annual Report to the President and Congress Fiscal Year 1983* (Office of the Secretary of Defense), pp. 1–91.
20 Paul F. Herman, Jr, "The Revolution in 'Military' Affairs," *Strategic Review* (Spring 1996), p. 27.
21 Derek Leebaert, p. 509.

22 Derek Leebaert, p. 512.
23 Commission co-chairmen Fred C. Ikle and Albert Wohlsetter memorandum entitled "Discriminate Deterrence" submitted to the Secretary of Defense and National Security Advisor (Washington, DC: Department of Defense, 1988), p. 8.
24 Richard H. Van Atta, Sidney G. Reed, and Seymour J. Deitchman, *DARPA Technical Accomplishments Vol. II: An Historical Review of Selected DARPA Projects* (Alexandria, VA: Institute for Defense Analyses, 1991), pp. 7–8.
25 Ralph Sanders, "Introduction" in Franklin D. Margiotta and Ralph Sanders (eds) *Technology, Strategy and National Security* (Washington, DC: National Defense University Press, 1985), pp. 4–5.
26 Ibid., p. 252.
27 Derek Leebaert, p. 515.
28 Major General Robert H. Scales, Jr (Retired), *Yellow Smoke: The Future of Land Warfare for America's Military* (New York: Rowman & Littlefield Publishers, Inc., 2003), p. 2.
29 Scales, p. 2.
30 Richard Van Atta and Michael J. Lippitz, *Transformation and Transition: DARPA's Role in Fostering and Emerging Revolution in Military Affairs, Part I: Overall Assessment* (Alexandria, VA: Institute for Defense Analyses, April 2003), pp. 5–6. Original in italics.
31 General Alexander Haig's address to the US Army Association October 13, 1976 reproduced in "Documentation" section, *Survival* (Vol. 19, No. 1), p. 34. Emphasis added.
32 Among the earliest observers of the OMG was Christopher N. Donnelly. His "The Soviet Operational Maneuver Group: A New Challenge for NATO," *International Defense Review* (September 1982), pp. 1177–1186 provided the first unclassified extended treatment of the OMG threat to NATO.
33 Lieutenant General James King, interview with author, August 12, 2003.
34 F. Digby and G. K. Smith, *Background on PGMs for NATO: Summarizing our Quick Look* (Santa Monica, CA: RAND, December 1973), p. 5.
35 Richard M. Swain, "AirLand Battle" in George F. Hofmann and Donn A. Starry (eds), *Camp Colt to Desert Storm: The History of U.S. Armored Forces* (Lexington, KN: University of Kentucky Press, 1999), p. 383
36 Ibid., p. 147.
37 Testimony of General Donn Starry, House of Representatives, *NATO Conventional Capability Improvement Initiatives*, Committee on Armed Services, Procurement and Military Nuclear Systems Subcommittee Jointly With the Research and Development Subcommittee (Washington, DC: Congressional Record, April 25, 1983), pp. 1841–1842.
38 Jacques Gansler, "The U.S. Technology Base: Problems and Prospects" in Franklin D. Margiotta and Ralph Sanders (eds) *Technology, Strategy and National Security* (Washington, DC: National Defense University Press, 1985), p. 105.
39 Gansler interview, University of Maryland, College Park, December 2, 2003.
40 Richard M. Swain, "AirLandBattle" in George F. Hofmann and Donn A. Starry (eds), *Camp Colt to Desert Storm: The History of U.S. Armored Forces* (Lexington, KN: University of Kentucky Press, 1999), p. 381.
41 Edward C. Meyer and R. Manny Ancell, *Who Will Lead? Senior Leadership in the United States Army* (Westport, Connecticut: Praeger, 1995), pp. 163–164.
42 Ibid., p. 182.
43 Edward N. Luttwak, "The Operational Level of War," *International Security* (Winter 1980/1981), p. 61. Cited in Lambeth, p. 81.
44 Ibid.
45 Colonel Harry G. Summer, Jr, *On Strategy: The Vietnam War in Context* (Carlisle Barracks, PA: Strategic Studies Institute, 1982).

46 Kenneth Allard, *Command, Control and the Common Defense* (Washington, DC: 1990), p. 185.
47 Allard, p. 185.
48 British Atlantic Committee, *Diminishing the Nuclear Threat: NATO's Defense and New Technology* (London: British Atlantic Committee, 1984).
49 Bernard W. Rogers, "Follow-on Forces Attack (FOFA): Myths and Realities" in *NATO Review* (December 1984), p. 8.
50 Allard, p. 185.
51 Interview, Colonel (Retired) Richard Johnson, PhD, July 21, 2001.
52 Cecil V. Crabb, Jr, *The Doctrines of American Foreign Policy: Their Meaning, Role and Future* (Baton Rouge, LA: Louisiana State University Press, 1982), p. 349.
53 Kenneth N. Waltz, "A Strategy for the Rapid Deployment Force," *International Security* (Vol. 5, No. 4), pp. 49–50. pp. 52–53. Richard Holloran, "Brown Warns That a Persian Gulf War Could Spread," *New York Times* (February 15, 1980), p. A3.
54 W. Scott Thompson, "The Persian Gulf and the Correlation of Forces" in *International Security* (Vol. 7, No. 1), p. 172. [pp. 157–180].
55 Thompson, p. 173.
56 NSDD cited Joseph T. Stanik, *El Dorado Canyon: Reagan's Undeclared War With Qaddafi* (Annapolis, MD: Naval Institute Press, 2003), p. 93.
57 Mark Huband, *Warriors of the Prophet: The Struggle for Islam* (Boulder, CO: Westview Press, 1999), p. 2.
58 CIA estimate cited in Joseph T. Stanik, *El Dorado Canyon: Reagan's Undeclared War With Qaddafi* (Annapolis, MD: Naval Institute Press, 2003), pp. 98–99.
59 Carl H. Builder, *Strategic Conflict Without Nuclear Weapons* (Santa Monica, CA: RAND, 1983), p. v.
60 Ibid.
61 Marshall quoted in Watts and Murray, "Military Innovation in Peacetime," p. 377.
62 Christian Nunlist, *Cold War Generals: The Warsaw Pact Committee of Defense Ministers, 1969–90*, Parallel History Project on NATO and the Warsaw Pact, May 2001. pp. 14–15.
63 Marshall in Krepinevich (2002), p. i.
64 Warner quoted in the transcript of the 1999 Fletcher Conference, panel 5, "Redefining Defense: Preparing U.S. Forces for the Future" (November 3, 1999), p. 11.
65 Owens, *Lifting the Fog of War*, p. 83.
66 Charles E. Heller and William A. Stofft, *America's First Battles, 1776–1965* (Lawrence, KS: University Press of Kansas, 1986), p. xiii.
67 General Edward C. Meyer, "Low-Level Conflict: An Overview" in Brian M. Jenkins (ed.), *Terrorism and Beyond: An International Conference on Terrorism and Low-Level Conflict* (Santa Monica, CA: The RAND Corporation, 1982), p. 39. See also Steven Metz, "A Flame Kept Burning: Counterinsurgency Support After the Cold War," *Parameters* (Autumn 1995), pp. 31–41. Metz is also online at http://carlisle-www.army.mil/usawc/Parameters/1995/metz.htm
68 Eliot A. Cohen, "Constraints on America's Conduct of Small Wars" in *International Security* (Vol. 9, No. 2), p. 180.
69 Generals Decker and Taylor, as quoted by Michael Lind, *Vietnam, The Necessary War* (New York: The Free Press, 1999), p. 103.
70 Cohen, p. 180.
71 Lieutenant General Wallace H. Nutting quoted in *Newsweek* (June 6, 1983), p. 24; Cited in Cohen, p. 181.
72 Metz, p. 6. The field manual was published in 1995.
73 Wolfowitz cited in James Mann, *Rise of the Vulcans: The History of Bush's War Cabinet* (New York: Viking Publishing, 2004), p. 360.
74 Cohen, p. 180.

75 Andrew F. Krepinevich and Steven M. Kosiak, "Smarter Bombs, Fewer Nukes," *The Bulletin of Atomic Scientists* (Vol. 54, No. 6), p. 8.

6 From RMAs to transformation: rediscovering the innovation imperative

 1 Stephen Biddle, "Victory Misunderstood: What the Gulf War Tells Us about the Future of Conflict," *International Security* (Vol. 21, No. 2), p. 142.
 2 Kenneth Allard, *Command, Control and the Common Defense* (Washington, DC: 1990), p. 274.
 3 Aston B. Carter and William J. Perry, *Preventive Defense: A New Security Strategy for America* (Washington, DC: Brookings Institute Press, 1999), p. 180.
 4 Edward N. Luttwak, *Strategy: The Logic of War and Peace* (Cambridge, MA: The Belknap Press of Harvard University Press, 2001), p. 94.
 5 Benjamin S. Lambeth, *The transformation of American Air Power* (Ithaca, NY: Cornell university Press, 2000), p. 150.
 6 Ibid., p. 122.
 7 Ibid.
 8 Ibid., p. 123.
 9 Vernon Loeb, "Bursts of Brilliance," *Washington Post Magazine* (December 12, 2002), p. 8.
10 Allan Millet, Williamson Murray, and Kenneth Watman, "The Effectiveness of Military Organizations" in Allan R. Millet and Williamson Murray, *Military Effectiveness, Volume I: The First World War* (Boston, MA: Unwin Hyman, 1988), p. 2.
11 Andrew Marshall, "Forward" to Andrew F. Krepinevich, Jr, *The Military-Technical Revolution: A Preliminary Assessment* (Washington, DC: Center for Strategic and Budgetary Assessments, 2002), pp. i–ii. The original Net Assessment was issued in July of 1992. The 2002 publication includes new forwards by Andrew Marshall and Andrew Krepinevich.
12 Andrew F. Krepinevich, Jr, *The Military-Technical Revolution: A Preliminary Assessment* (Washington, DC: Center for Strategic and Budgetary Assessments, 2002), p. 8.
13 William J. Perry, "Military Action: When to Use and How to Ensure its Effectiveness" in James E. Nolan (ed.), *Global Engagement: Cooperation and Security in the 21st Century* (Washington, DC: Brookings Institution Press, 1994), p. 240.
14 See, for example, then Deputy Director of Defense Research and Engineering Frank Kendell, "Exploiting the Military Technical Revolution: A Concept for Joint Warfare" in *Strategic Review* (Spring 1992), pp. 23–30. Among the large literature on RMAs are Williamson Murray, "Thinking About Revolutions in Military Affairs," *Joint Force Quarterly* (Summer 1997), Robert Tomes, "Revolution in Military Affairs—A History," *Military Review* (September/October 2000), pp. 98–101; Lawrence Freedman, *The Revolution in Military Affairs*, Adelphi Paper 318 (IISS 1998); and Richard O. Hundley, *Past Revolutions, Future Transformations: What Can the History of Revolutions in Military Affairs Tell Us About Transforming the U.S. Military?* (Washington, DC: RAND, 1999). Later chapters provide a more extensive discussion on RMAs.
15 Abhi Shelat, "An Empty Revolution: MTR Expectations Fall Short," Harvard International Review (Summer 1994), p. 52.
16 James Der Derian, *Virtuous War: Mapping the Military-Industrial-Media-Entertainment Network* (Boulder, CO: Westview Press, 2001), p. 28; Andrew Marshall, Director of Net Assessment (Office of the Secretary of Defense), *Memorandum for the Record*, "Some Thoughts on Military Revolutions—Second Version," August 23, 1993.
17 Der Derian, pp. 28–29.

18 Steven Metz and James Kievit, *Strategy and the Revolution in Military Affairs: From Theory to Policy* (Carlisle Barracks, PA: U.S. Army War College, 1995), 1.

19 Attributed to Dennis Showalter's paper presentation, "The Wars of Moltke, an RMA" presented at the Revolution in Military Affairs Conference, Marine Corps Combat Development Command, Quantico, VA, April 1996. Quoted in Williamson Murray, "Introduction," in Williamson Murray (ed.), *The Emerging Strategic Environment* (Westport, CT: Praeger, 1999), p. xxvii.

20 Krepinevich, p. 30.

21 Paul K. Van Riper and F. G. Hoffman (1998) "Pursuing the Real Revolution in Military Affairs: Exploiting Knowledge-Based Warfare," *National Security Studies Quarterly*, Vol. IV, No. 3 (Summer 1998), p. 2.

22 James Fitz-simonds and Jan van Tol, p. 25.

23 Hundley, *Past Revolutions, Future Transformations* (Santa Monica, CA: RAND, 1999), p. 9.

24 Williamson Murray and MacGregor Knox, "Thinking About Revolutions in Warfare" in MacGregor Knox and Williamson Murray (eds) *The Dynamics of Military Revolution, 1300–2050* (New York: Cambridge University Press, 2001), p. 1.

25 Stephen Biddle, *The RMA and the Evidence: Assessing Theories of Future Warfare*, Institute for Defense Analyses, manuscript dated August 8, 1996.

26 Francis Fukuyama, *The End of History and the Last Man* (New York: The Free Press, 1992); see also Timothy Burns (ed.), *After History? Francis Fukuyama and His Critics* (London: Rowman and Littlefield, 1994).

27 Martin Van Creveld, *The Transformation of War* (New York: The Free Press, 1991).

28 The prototypical work of the new mood in governance was David Osborne and Ted Graebler, *Reinventing Government: How the Entrepreneurial Spirit is Transforming the Public Sector* (New York: Penguin, 1993).

29 Colin S. Gray, *The American Revolution in Military Affairs: An Interim Assessment* (Strategic and Combat Studies Institute Occasional Paper 28, 1997), fn 1, p. 5. See also Metz and Kievit, p. 2.

30 Gray, *Strategy For Chaos*, p. xiii.

31 Michale O'Hanlon, *Technological Change and the Future of Warfare* (Washington, DC: Brookings Institute Press, 2000), p. 5.

32 Ibid., p. 31.

33 Andrew W. Marshall, Director for Net Assessment, Office of the Secretary of Defense, testifying before Senate Armed Services Committee, Subcommittee on Acquisition and Technology, May 5, 1995, p. 1.

34 National Defense Panel, *Transforming Defense: National Security in the 21st Century* (December 1997), p. 51.

35 Paul H. Nitze, "A Threat to Ourselves," *New York Times* (October 28, 1999), p. 25A.

36 Andrew F. Krepinevich and Steven M. Kosiak, "Smarter Bombs, Fewer Nukes," *The Bulletin of Atomic Scientists* (Vol. 54, No. 6), p. 7.

37 John B. Hattendorf, *The Evolution of the U.S. Navy's Maritime Strategy, 1977–1986* (Newport, RI: Center for Naval War College Press, Newport Paper 19, 2004), p. 4.

38 Williamson Murray and MacGregor Knox, "Thinking About Revolutions in Warfare" in MacGregor Knox and Williamson Murray (eds), *The Dynamics of Military Revolution, 1300–2050* (New York: Cambridge University Press, 2001), p. 1.

39 William A. Owens, "Creating a U.S. Military Revolution" in Theo Farrell and Terry Terriff (eds), *The Sources of Military Change: Culture, Politics, Technology* (Boulder, CO: Lynnee Rienner, 2002), p. 209.

40 Ibid., p. 211.

41 Ibid., p. 209.

42 Douglas A. Macgregor, "The Joint Force: A Decade, No Progress," *Joint Force Quarterly* (Winter 2000–2001), p., 20. [18–23].

43 Janne E. Nolan, *An Elusive Consensus: Nuclear Weapons and American Security After the Cold War* (Washington, DC: Brookings Institution Press, 1999), p. 86.

44 Nolan, *An Elusive Consensus*, p. 86.

45 Robert W. Gaskin, "A Revolution for the Millennium" in Williamson Murray (ed.), *The Emerging Strategic Environment: Challenges of the Twenty-First Century* (Westport, CT: Praeger, 1999), p. 133.

46 Ibid.

47 Ibid., p. 153.

48 Ibid., p. 56.

49 Ibid., p. 41.

50 Gray, *Strategy For Chaos*, p. 17.

51 Van Creveld, (New York: The Free Press, 1991).

52 *Annual Report to the President and Congress* (Washington, DC: Department of Defense, 1999), chapters 13–15.

53 Ibid.

54 Ted Gold, *Report of the Defense Science Board Task Force on DoD Warfighting Transformation* (Washington, DC: Office of the Undersecretary of Defense, 1999), p. 3

55 Ibid.

56 Ibid.

57 Ibid., p. 30.

58 Ibid., p. 1.

59 Eliot A. Cohen, "Defending America in the Twenty-First Century," *Foreign Affairs* (November/December 2004), p. 41, [40–56].

60 Ibid., p. 41.

61 Ashton B. Carter, "Keeping America's Military Edge" in *Foreign Affairs* (January/February 2001), p. 92.

62 Andrew J. Bacevich, *American Empire: The Realities and Consequences of U.S. Diplomacy* (Cambridge, MA: Harvard University Press, 2002), p. 140.

63 Department of Defense Transformation Study Group, "Transforming Military Operational Capabilities" (www.defenselink.mil.news.Nov20001/t11272001_t1127 ceb.html), p. 5.

64 Bush cited in Bacevich, p. 140.

65 Ibid.

66 Quoted in Tom Philpott, "New 'Transformation Chief' Says 9-11 Should Shake Status Quo," *Newport News Daily Press* (November 30, 2001).

67 Donald H. Rumsfeld, "Beyond This War on Terrorism," *Washington Post* (November 1, 2002), p. 35.

68 Quoted in Adam J. Hebert, "Aldridge: War on Terrorism Demands Major Changes in Acquisition Practices," *InsideDefense.com* (October 31, 2001).

69 See, for example, Ann Roosevelt, "Chief Scientist: Army to Accelerate FCS Technologies," *Defense Week* (November 13, 2001), p. 1; Greg Jaffe and Anne Marie Squeo, "High-Tech Eyes, Ears Face Battle With Means of Traditional Warfare," *Wall Street Journal* (September 19, 2001), p. 1.

70 *Annual Report to the President and Congress* (Washington, DC, Department of Defense, 2001), p. 4.

71 David L. Norquist, "The Defense Budget: Is it Transformational?" in *Joint Forces Quarterly* (Summer 2002), p. 94. [pp. 91–99].

72 Department of Defense Briefing, "Findings of the Nuclear Posture Review (January 9, 2002).

73 Krepinevich cited in Vernon Loeb, "Billions, and it Can't Make Change," *Washington Post* (September 13, 2002), p. A37.

74 Andrew F. Krepinevich, Jr, *The Military-Technical Revolution: A Preliminary Assessment* (Washington, DC: Center for Strategic and Budgetary Assessments, 2002), p. 22.

75 Ibid., p. 22.
76 Kevin R. Cunningham, *The Changing Relationship Between Intelligence and Strategy: Paradoxes and Possibilities* (Carlisle Barracks, PA: U.S. Army War College, 2001), p. 19.
77 David Harvey, *The Condition of Postmodernity* (Oxford: Basil Blackwell, 1989), p. 147.
78 Kendall, *Strategic Review*, p. 25.
79 *Foreign Affairs* (Vol. 75, No. 2, 1996), p. 20.
80 Norman C. Davis, "An Information-Based Revolution in Military Affairs" in John Arquilla and Davis Ronfeldt (eds), *In Athena's Camp: Preparing for Conflict in the Information Age* (Washington, DC: RAND, 1997), p. 83.
81 Scales, p. 3.
82 See Robert R. Tomes, "Boon or Threat? Information Warfare and U.S. National Security," *Naval War College Review* (Summer 2000), p. 39, [39–59].
83 Douglas A. Macgregor, *Breaking the Phalanx: A New Design for Landpower in the 21st Century* (Westport, CT: Praeger, 1997), p. 45.
84 Joseph S. Nye, *The Paradox of American Power: Why the World's Only Superpower Can't Go it Alone* (New York: Oxford University Press, 2002), p. 67.
85 Cunningham, p. 18.
86 "Bush: West Point Grads Answer History's Call to Duty," *Point View* (June 7, 2002) at www.usma.edu/PublicAffairs/PV/CallToDuty.htm; Fred Barnes, "Bush's Big Speech," *The Weekly Standard* (June 17, 2002).
87 Bob Woodward, *Plan of Attack* (New York: Simon and Schuster, 2004), p. 131.
88 David Cannadine (ed.), *The Speeches of Winston Churchill* (London: Penguin Books, 1989), p. 303.
89 David Cannadine (ed.), *The Speeches of Winston Churchill* (London: Penguin Books, 1989), pp. 296–297.
90 *National Security Strategy of the United States.* Washington, DC, September 2002 (www.whitehouse.gov/nsc/nss.pdf), p. 15.
91 *Military Transformation: A Strategic Approach*, (Washington, DC: Director, Force Transformation, Office of the Secretary of Defense, Fall 2003), p. 8.
92 John A. Lynn, "Reflections on the History and Theory of Military Innovation and Diffusion" in Colin Elman and Miriam Fendius Elman (eds), *Bridges and Boundaries: Historians, Political Scientists, and the Study of International Relations* (Cambridge, MA: The MIT Press, 2001), p. 359.
93 Ibid., p. 360.
94 *Military Transformation: A Strategic Approach*, (Washington, DC: Director, Force Transformation, Office of the Secretary of Defense, Fall 2003), p. 6.
95 On sustaining versus disruptive change, see Clayton M. Christensen, *The Innovator's Dilemma: When New Technologies Cause Great Firms to Fail* (Boston, MA: Harvard Business School Press, 1997).
96 Cited in Frances Cairncross, *The Death of Distance* (Boston, MA: Harvard Business School Press, 1997), p. 3.
97 Harvey Brooks, "Technology, Evolution, and Purpose" in Stephen R. Graubard (ed.), *Modern Technology: Problem of Opportunity?* (Daedalus Vol. 109, No. 1), pp. 65–66.

7 Conclusion: revisiting the military innovation framework

1 *What is History?* (New York: Vintage Books, 1961), pp. 125–126.
2 Perry in Ashton B. Carter and William J. Perry, *Preventative Defense: A New Security Strategy for America* (Washington, DC: Brookings Institution Press, 1999), p. 178.
3 Marshall in Krepinevich (2002), p. i.
4 Christian Nunlist, *Cold War Generals: The Warsaw Pact Committee of Defense Ministers, 1969–90*, Parallel History Project on NATO and the Warsaw Pact, May 2001, pp. 14–15.

5 Allan R. Millet, "Patterns of Military Innovation in the Interwar Period" in Williamson Murray and Allan R. Millett (eds), *Military Innovation in the Interwar Period* (New York: Cambridge University Press, 1996), p. 335.

6 Rosen, p. 251.

7 Barry Watts and Williamson Murray, "Military Innovation in Peacetime" in Williamson Murray and Allan R. Millett (eds), *Military Innovation in the Interwar Period* (New York: Cambridge University Press, 1996), p. 371.

8 Williamson Murray, "Innovation Past and Future" in Murray and Millett (eds), *Military Innovation in the Interwar Period*, p. 300.

9 U.S. Congress, Office of Technology Assessment, *Technologies for NATO's Follow-On Forces Attack Concept* (Washington, DC: Government Printing Office, July 1986), p. 7.

10 Ibid.

11 Frank Kendall, "Exploiting the Military Technical Revolution: A Concept for Joint Warfare" in *Strategic Review* (Spring 1992), p. 24.

12 Richard Van Atta, *Transformation and Transition: DARPA's Role in Fostering an Emerging Revolution in Military Affairs*, Vol. I: Overall Assessment (Alexandria, VA: Institute for Defense Analyses, April 2003), p. 10.

13 Derek Leebaert, *The Fifty-Year Wound: The True Price of America's Cold War Victory* (Boston, MA: Little, Brown and Company, 2002), p. 584.

14 Owens, *Lifting the Fog of War* (New York: Farrar, Straus, Giroux, 2000), p. 52.

15 Interview with MG Stan McChrystal, Pentagon, September 9, 2003.

16 Defense Intelligence Agency, *Collection/C4ISR Support to Targeting and Precision Strike* (August 1998), pp. 18; xvii.

17 Vernon Loeb, "Bursts of Brilliance," *Washington Post Magazine* (December 12, 2002), p. 8.

18 Thomas P. Hughes, *Rescuing Prometheus: Four Monumental Projects that Changed the Modern World* (New York: Vintage Books, 1998), p. 96. I'm indebted to Richard Van Atta for reminding me of the role Schriever played in the evolution of systems engineering as an American discipline.

19 Hughes, p. 119.

20 Hughes, p. 4. Hughes provides a detailed history of SAGE and Atlas.

21 Michael Hobday, Andrea Prencipe, and Andrew Davies, "Introduction" in Andrea Prencipe, Andrew Davies, Michael Hobday (eds), *The Business of Systems Integration* (New York: Oxford University Press, 2003), p. 16.

22 Henry D. Levine, "Some Things to All People: The Politics of Cruise Missile Development," *Public Policy* (Vol. 25, No. 1), p. 124, [117–168].

23 Ibid., p. 125.

24 Ibid., p. 122, [117–168].

25 Williamson Murray, "Afterward" in Williamson Murray (ed.), *The Emerging Strategic Environment: Challenges of the Twenty-First Century* (Westport, CT: Praeger, 1999), p. 269. Emphasis in original.

26 Nicholson, p. 236.

27 Ibid., p. 236.

28 See Lawrence B. Mohr, *Explaining Organizational Behavior: The Limits and Possibilities of Theory and Research* (San Francisco, CA: Jossey-Bass, 1982).

29 John Lewis Gaddis, *The United States and the End of the Cold War: Implications, Reconstructions, Provocations* (New York: Oxford University Press, 1992), p. 3.

Index

CPSIA information can be obtained at www.ICGtesting.com
Printed in the USA
LVOW07s1413210813

348979LV00002B/22/A